THE RED ATLAS

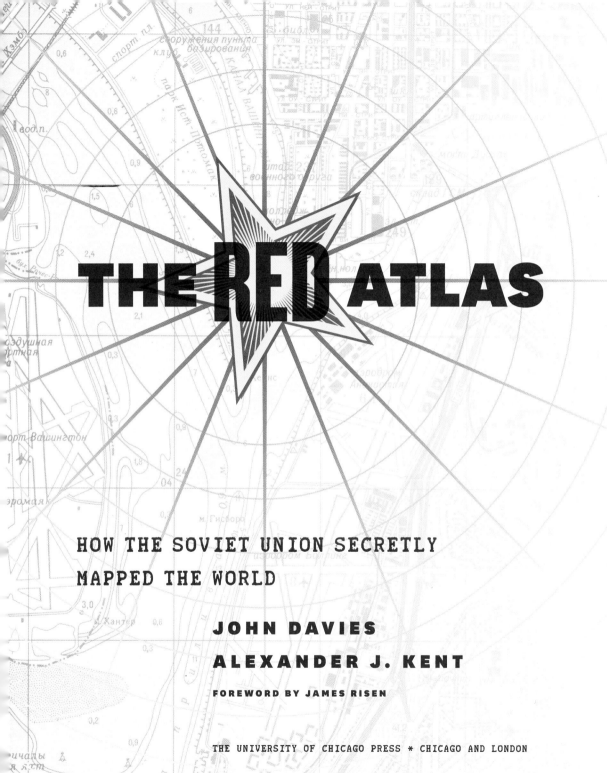

THE RED ATLAS

HOW THE SOVIET UNION SECRETLY MAPPED THE WORLD

JOHN DAVIES
ALEXANDER J. KENT

FOREWORD BY JAMES RISEN

THE UNIVERSITY OF CHICAGO PRESS ∗ CHICAGO AND LONDON

The University of Chicago Press, Chicago 60637
The University of Chicago Press, Ltd., London
© 2017 by The University of Chicago
Foreword © 2017 by James Risen
All rights reserved. No part of this book may be used or reproduced in any manner whatsoever without written permission, except in the case of brief quotations in critical articles and reviews. For more information, contact the University of Chicago Press, 1427 E. 60th St., Chicago, IL 60637.
Published 2017
Printed in Canada

26 25 24 23 22 21 20 19 18 17   2 3 4 5

ISBN-13: 978-0-226-38957-8 (cloth)
ISBN-13: 978-0-226-38960-8 (e-book)
DOI: 10.7208/chicago/9780226389608.001.0001

LIBRARY OF CONGRESS CONTROL NUMBER: 2017007292

♾ This paper meets the requirements of ANSI/NISO Z39.48-1992 (Permanence of Paper).

Dedicated to the thousands of men and women who
created the treasure trove of maps described in this book
and
to our next generation,
Abigail, Edward, and Sophia,
in the hope that they inherit a more harmonious world

CONTENTS

FOREWORD BY JAMES RISEN IX
NOTE TO READERS XIII

INTRODUCTION
Why this book is a detective story 1

1 WAR AND PEACE
The background of the story—from Napoléon's march on Moscow to the collapse of the Soviet Union 3

2 CAPTURING THE WORLD—ON PAPER
Describing the style, content, and symbology of the Red Army's maps of the world 9

3 PLOTS AND PLANS
The overt and covert methods of the Soviet cartographers 47

4 RESURRECTION
The discovery of the maps after the fall of the Soviet Union and their continuing significance today 131

ACKNOWLEDGMENTS 145

APPENDIX 1	Examples of Maps of Various Series and Scales 146
APPENDIX 2	References and Resources 205
APPENDIX 3	Translation of Typical City Plan "Spravka" 211
APPENDIX 4	Translation of Typical Topographic Map "Spravka" 215
APPENDIX 5	Symbols and Annotation 219
APPENDIX 6	Glossary of Common Terms and Abbreviations 222
APPENDIX 7	Print Codes 223
APPENDIX 8	Secrecy and Control 225
INDEX	Place-Names 227
INDEX	General 231

FOREWORD
BY JAMES RISEN

NEARLY THREE DECADES AFTER THE FALL OF THE BERLIN WALL, Cold War secrets are still tumbling out.

Some of them are beautiful.

An enormous and secret infrastructure supported the intelligence battles that were waged between the East and West throughout the forty-year Cold War standoff. The United States and the Soviet Union and their allies spied on each other incessantly, because they wanted to be prepared just in case an unthinkable war ever broke out.

Spying involves waiting, watching, remembering, and recording. It involves sophisticated cameras and high-altitude aircraft and missiles with satellites—but also people on the ground, quietly walking down streets, looking.

Sometimes, the products of all that spying during the Cold War were intelligence reports, which told Washington or London or Moscow what the other side was doing. Intelligence reports might provide inside information that could be used to decide broad strategy—when to move armies and navies.

Sometimes, the products were maps. Highly detailed maps, useful for spies and policy makers, for diplomats, invading armies, and occupiers. Maps that could provide more specific tactical information than might come from intelligence reports. Maps that could tell a general which roads and bridges provided the best routes to use to drive his tanks, or an admiral which harbors were deep enough for his destroyers.

When the Cold War ended, the secret infrastructure built up by the superpowers was left behind. Today much of it has been rediscovered and repurposed; missile silos in the American Midwest, for example, are being turned into eclectic prairie homes.

Now, with the publication of *The Red Atlas*, we discover another aspect of that long-secret infrastructure—Soviet military maps, including maps of Moscow's great adversaries, the United States and Britain.

Once classified, the maps displayed here give an eerie reminder of an obvious, yet still unsettling fact, at least for American and British readers. They show that the Russians were watching America and Britain, just as much as the Americans and British were watching them. They were looking down from above, and looking from the street. The Russians didn't miss much.

Americans have long taken for granted the fact that the United States had spy planes and spy satellites flying over the Soviet Union. But these maps remind us that the Soviets had spy satellites flying over the United States too, staring through the clouds at America as it changed and grew throughout the Cold War. In fact, Russian mapmakers worked hard to keep up with the transforming landscapes of the United States and Great Britain—the construction of interstate highways and shopping centers and new suburbs and new military bases posed endless challenges for the Russians. Yet the maps show that they were sometimes faster to incorporate new landmarks than were their Western counterparts.

What's more, the Russian maps sometimes included sensitive information about secret locations in the United States and Britain that had been excluded from public Western maps.

But the Russian maps include details that could not have come from satellites alone. They incorporate the names of factories and what was made in those factories. The kind of products manufactured inside a factory can't be divined from overhead imagery, so did the purpose of the factories become clear only when a Soviet agent walked down the street? Or did the Russians have other information? Were there moles in the U.S. and British governments passing them data? The questions remain unanswered.

These Soviet maps, which were produced as part of the world's largest mapping effort, present an alternative view of the globe. They tell us what the Russians saw when they looked at us. What the Russian mapmakers considered important enough to detail in their maps reveals to us what the Russians thought was most important in the United States and Britain.

These maps were the product of an ambitious effort by Moscow to accurately and secretly map the Soviet Union, its Eastern European allies, its Western adversaries—and the rest of the world. Conducted by the Military Topographic Directorate of the General Staff of the Soviet Army, the worldwide mapping effort may have produced more than a million different maps of different parts of the globe.

The Russian maps were also of extraordinarily high quality. Soviet-era military maps were so good that when the United States first invaded Afghanistan in late 2001, American pilots relied on old Russian maps of Afghanistan. For almost a month after the United States began a bombing campaign to help oust the Taliban government, American pilots were guided by Russian maps dating back to the Soviet occupation of Afghanistan in the 1980s. The U.S. intelligence community had been too slow in figuring out how to process and distribute up-to-date satellite photographs of Afghanistan.

These Russian military maps are detailed and honest, even in their careful depictions of the Soviet Union itself. That's ironic, since Russian-made maps of the Soviet Union that were made available to the public during the Cold War were purposefully terrible and misleading. A product of Stalinist paranoia of foreign invaders, the maps of the Soviet Union made for the public and tourists included flaws and mistakes designed to hide information or deflect travelers.

But the Russian military maps are also something else—long-lost works of art. The craftsmanship and the sheer beauty of the maps make them mesmerizing. The use of colors, lines, and geometric shapes lends them an art deco feel.

They don't appear to be the thoughtless products of a giant military enterprise. Instead, the maps have an artisanal quality. The careful dedication to detail—portraying factory building facades, roads and bridges, landscapes with forests that include individual trees—allows you to think that the mapmakers were thinking about more than just providing Soviet military officers with maps for marching.

The fact that cities, towns, and other landmarks on the maps of the United States and Britain are identified in the Cyrillic alphabet of the Russian language adds to the mystique. We are looking at our own homes through the artistry of our adversaries. It is strange to see Russian maps of New York, Chicago,

Los Angeles, London, and other Western cities; we can imagine that these would have been the official maps of the Soviet occupation of the United States and Great Britain if Moscow had ever invaded and won a war.

Of course, in a world of smartphones with GPS and driving apps with voices that tell us which roads to take, and which warn us where there is a traffic jam or a speed camera, paper maps may seem anachronistic.

But their historical significance cannot be denied. And neither can their beauty.

NOTE TO READERS

THE MAP EXTRACTS IN THIS BOOK ARE FROM MAPS IN PRIVATE collections. As the paper sheets are up to fifty years old, the quality of reproduction varies, for which we ask the reader's understanding.

REFERENCES IN SQUARE BRACKETS THROUGHOUT THE BOOK ARE to items listed in appendix 2. Supplementary information, links, and map images are at http://redatlasbook.com.

INTRODUCTION WHY THIS BOOK IS A DETECTIVE STORY

THIS IS A STORY THAT CAN ONLY BE TOLD BY THOSE WHO WERE not involved. It's the story of a massive secret project, started by Stalin, spanning fifty years and involving thousands of people—all sworn to secrecy. It's the story of the world's largest mapping endeavor and, arguably, the world's most intriguing maps. Even today, long after the end of the Cold War, the maps are classified as "Secret" in Russia; the people who worked on them remain silent and many of the maps remain hidden. The story of this amazing enterprise has never been told, and the maps themselves have rarely been publicly displayed. The secrecy has, however, been partially breached in the three Baltic states where the sudden collapse of the Soviet Union left a stockpile of maps in hastily abandoned depots in these newly independent countries. Even there, though, firsthand testimony is hard to find.

We, the authors of this book, are British map enthusiasts who have spent many years gathering a huge collection of these sheets during our travels, and we have diligently examined them in detail. This book is written for the general reader and anyone interested in the history and political geography of the twentieth century. We describe the scope and scale of the global mapping project, as much as can be deduced from the maps so far discovered; we readily acknowledge that we "don't know what we don't know." It is quite likely that many more maps remain undiscovered, and as they emerge (if they do), then the story will continue to develop.

From the evidence of what appears on the maps—what is shown, what omitted; what is correct, what erroneous—we try to deduce how the maps were made; how it was possible during the dark days of mutual suspicion and under the ever-present threat of mutual nuclear destruction for Soviet cartographers to collect such an astonishing wealth of detail about the streets, buildings, industries, transport, and utilities of capitalist cities. Like good detectives, we lay out the evidence and state our conclusions. We avoid speculation about unknowables and leave it to others to offer theories about the purpose of this vast enterprise.

THE BOOK HAS FOUR CHAPTERS AND EIGHT APPENDICES, WITH supplementary information and examples of maps on the accompanying website: http://redatlasbook.com.

Chapter 1 looks at the prehistory: how Russian cartographic expertise originated in tsarist days. The next two chapters focus on the Soviet military mapping of the Cold War era. Chapter 2 describes the maps as artifacts, defining the various series, scales, and specifications. In chapter 3 the question of how the maps were made is examined, with examples of interpretation and misinterpretation by the compilers of the data they had collected. Chapter 4 concerns post-Soviet times, after the fall of the Iron Curtain, and tells of the emergence of the maps in the West: their "afterlife" as the only reliable maps of many parts of the world, and their legacy long after the demise of the regime that produced them.

Appendix 1 includes a selection of examples of Soviet maps to showcase the range of styles and the evolution of the specifications over time. Other appendices provide a resource for assisting the interpretation of these masterpieces of cartography.

WAR AND PEACE

WHEREVER YOU ARE ON THE PLANET READING THIS BOOK, THE place is likely to have been mapped in detail by the Soviet Union. At least once. During the Second World War, Stalin ordered his military to conduct a world mapping program that was upheld by his successors throughout the Cold War. From their drawing offices around the USSR, thousands of Soviet cartographers were busy mapping the globe, from the pyramids to the Pentagon—and their tremendous legacy is only just coming to light.

Soviet maps are comprehensively detailed and include a wealth of information that goes beyond that of any published national topographic map series. They can tell you about the height of a bridge above water, its dimensions, load capacity, and the main construction material. They can tell you the width of a river, the direction of its flow, its depth, and whether it has a viscous bed. They can tell you the type of trees in a forest, as well as their height, girth, and spacing. They can also tell you the name of a factory and what it produces. At each scale, Soviet topographic maps conform to a consistent specification, using a standard symbology, projection, and grid.

Soviet maps are as chilling as they are amazing. Chilling because they have the power to unnerve us when we see the city where we live or the landscape of our childhood presented in unfamiliar colors, symbols, and, most of all, in an alien language—their bold Cyrillic place-names both arrest and fascinate our gaze. That these maps were created as part of a secret military program enhances their intrigue. What did they know? Was my street mapped? How did they do it? What did they get wrong? Our aim in writing this book is to answer some of these questions by bringing details of this unparalleled cartographic feat to light.

A BRIEF HISTORY OF SOVIET MAPPING

From Tsar Alexander repelling Napoléon to Stalin facing Hitler, Russian leaders have depended on the maps produced by the Military Topographic Depot (and its successors) to plan their campaigns. These maps have also been essential for the economic development of the country. While the same may be true for other countries, the vast extent of continental Russia and the harshness of the terrain and weather led to the emergence of arguably the most talented pool of geodesists, topographers, surveyors, and cartographers the world has seen.

The first detailed map of the Russian Empire was produced by the St. Petersburg Map Depot in 1801. This was at the scale of 20 versts (21.3 km/13.25 miles) to the inch (1:840,000) and became known as the hundred-sheet map. The Russian Military Topographic Depot was founded in January 1812, five months before Napoléon and his Grande Armée crossed the river Neman during his ill-fated Russian campaign. By 1840 the 10-versts map (1:420,000) of much of European Russia was published. Illustrations of these and many other early Russian maps appear in Alexey Postnikov's *Russia in Maps* [28]. The focus on mapping European Russia, such as the Finnish and Polish territories, continued until the Second World War.

Modern topographic mapping of the Soviet Union began after the Bolshevik Revolution of 1917, with the first maps at 1:1,000,000 completed in 1918 using a new system of sheet lines based on the system proposed for Albrecht Penck's International Map of the World (IMW), which had been begun in 1913 (see fig. 2.10).

In 1919 Lenin's decree put all mapping activities and relevant control functions under the state's supervision. The year 1921 saw the introduction of a standard specification for military topographic maps at a range of scales (1:10,000, 1:25,000, 1:50,000, 1:100,000, 1:200,000, 1:500,000, 1:1,000,000) and the use of these sheet lines, with the first sheets of a metric series derived from photogrammetry (aerial photography) published in 1924. Maps at such a wide range of scales would be useful in supporting a full scope of activities, from planning regional strategies to reading local terrain and street-level detail.

The signing of the Molotov-Ribbentrop Pact (the Pact of Steel) earlier in 1939 had divided eastern European states into zones of influence: Poland was to be divided between Germany and the Soviet Union, while the Soviet Union was to invade the Baltic states and Finland. This cooperation between Hitler and Stalin achieved political and economic aims: a trade pact, the German-Soviet Commercial Agreement, was signed in 1940, yet despite suspicion of Hitler's motives, Stalin did not consider Nazi Germany to pose an immediate threat at the time.

Following successive revisions, the specification of 1940 incorporated the inclusion of higher levels of topographic detail than present in the national mapping of other countries, including types of forest and the widths of roads. The mobilization of Soviet forces in response to Operation Barbarossa, the Nazi invasion of the Soviet Union on June 22, 1941, brought greater focus on the compilation of topographic material in support of military operations but also provided an opportunity to advance the longer-term Soviet objective of achieving global communism.

From the Second World War, foreign towns and cities were mapped at larger scales, generally 1:10,000 and 1:25,000. These secret map series were also produced to a comprehensive specification, which ensured that the typefaces, colors, symbology, and projection system in use would be standardized whatever and wherever the city or town mapped. Curiously, the sheets have no fixed size or format, allowing the cartographers to determine the best solution for covering the geographical shape and general extent of the town or city. The standard specification of the plans seems to have evolved in stages, each offering a more sophisticated representation of the urban landscape and, in particular, a more detailed classification of strategically important buildings. Earlier plans employed fewer colors (i.e. light brown, dark brown, blue, and green), with important buildings shown in dark brown overprinted with blue to make them appear black and therefore visually more prominent on the plan.

The introduction of further printing plates can be appreciated on the plan of Belfast in Northern Ireland, printed in 1964, which uses different shades of brown to distinguish between prominent buildings and neighborhoods (or blocks) that are more or less densely built-up (fig. 1.1). By the early 1970s,

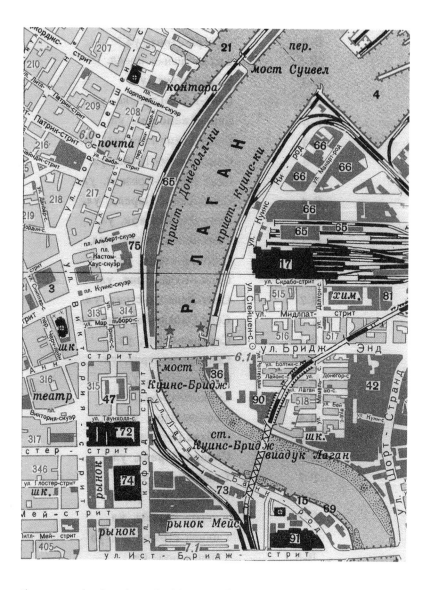

1.1 *Belfast*, 1:10,000, 1964.

the range of color plates had increased to ten (i.e., light blue, dark blue, light green, dark green, light yellow, orange, purple, gray, brown, and black), which facilitated an extension of the symbology and allowed the three major categories of important buildings (i.e., military and communications, governmental and administrative, and military-industrial) to be printed using different color plates.

Printing such large-format plans in so many colors with near-perfect print

1.2 A reentry vehicle from the Zenit satellite revealing the camera ports. Photograph: Maryanna Nesina.

registration itself testifies to the skill of the printers in the military map printing factories across the former Soviet Union. The quality of printing reflects the level of training and the reliability of humidity-control equipment and the electricity supply at the time.

While national mapping was undoubtedly used among other types of information sources, the successful launch of the Soviet program of Zenit satellites in 1962 saw an increased reliance on reconnaissance imagery (fig. 1.2). Yet the plans, especially, include information that would have been virtually impossible to derive from remote sensing, such as the names of factories and the products manufactured there, and indications of the load capacity of bridges, for example. The maps also include disused railways and old streetcar and ferry routes as well as detailed depictions of the terrain. Seeing the landscape through the eyes of a Soviet cartographer—whatever resource could be potentially useful, whatever information could be found—was added. Each map is therefore a formidable inventory of geospatial information.

Interestingly, the IMW relied upon the collaboration of national governments, and as a mapping project it ultimately failed, with less than half of the planned 2,500 sheets produced. By contrast, the product of the Soviet military mapping program is likely to run into millions of different sheets; the 13,133 sheets to cover the USSR alone at the scale of 1:100,000—ten times that of the IMW sheets—were completed by 1954.

Mapmaking involves choosing what to show and how to show it, and in contrast to the digital mapping available today, the paper map, as a static medium, requires every type of information to be shown simultaneously. Ensuring that all features are legible and clear involves tremendous skill in design and draftsmanship, particularly for detailed maps such as these. Yet these choices also reveal more about the values of the cartographers or, in this case, the state that produced them. The maps were used as an inventory of geospatial intelligence; each sheet is a rich topographic database that is a product of the laborious process of compiling material from a range of sources. Unlike any national topographic map series, they present a sum of past landscapes; they include traces of infrastructure that have long fallen into disuse. Very little, it would seem, was deemed insignificant or irrelevant to the Soviet eye.

Far from their original circumstances of production, what these maps offer today presents an altogether different resource. They offer a fascinating glimpse of the view from the other side of the Iron Curtain. Secrecy and fear have been replaced by discovery and fascination. Soviet maps demonstrate that the meaning of maps is never constant; there are always new ways in which a map can be used and can change the world.

CAPTURING THE WORLD—ON PAPER

MAPS ARE INSTRUMENTS OF POWER, AND STALIN'S DECISION TO invest further in their ability to facilitate the running of the state has bequeathed to the world the legacy of an unmatched geospatial resource. Stalin decreed that the first priority for the Military and Civil State Topographic Services after the war was to complete the survey of the entire territory of the Soviet Union for the 1:100,000 topographic map. This was to be based on aerial photogrammetry but would still rely upon field geodetic control.

This enormous task was achieved by 1954, and the resulting 1:100,000 survey was then used to derive the smaller-scale 1:200,000 and 1:500,000 maps. None of these, however, were available to the public. Maps for ordinary citizens were based on a 1:2,500,000 map of the country and, as such, were inadequate for any detailed use. Moreover, random distortions and inaccuracies were tossed in for good measure. These official maps, the only ones available to the general public and primarily designed for tourists, were published by the GUGK (Central Administration for Geodesy and Cartography of the USSR Council of Ministers—Главное Управле-ние Геодезии и Картографии при Совете Министров СССР) (see figs. 2.1 and 2.2). The GUGK was closely allied to the VTU (the Military Topographic Directorate of the General Staff of the Soviet Army—Военно-Топографическое управление Генерального Штаба Советской Армии), the sister military establishment.

In the 1970s, the need for detailed mapping for use by civil authorities led to the introduction of the so-called SK-63 Series, constructed on disparate systems of sheet lines and carrying no geodetic data but otherwise accurate.

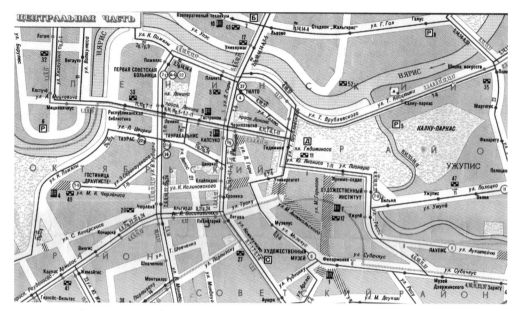

2.1 Enlarged city-center inset in official public map of Vilnius, Lithuania, published by the GUGK in 1981. No scale is given for the main plan or the inset. Compare this with the military version in figure 3.49. The reverse of the map has a list of local bus routes, a diagram of national bus and rail routes, a street index, and a list of tourist attractions.

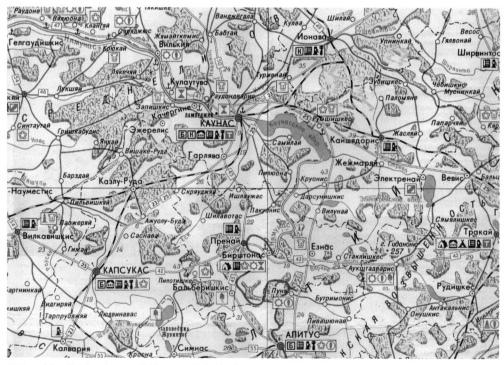

2.2 Part of the official tourist map of Lithuania, published by the GUGK in 1980. No scale is given. The reverse has descriptions and photographs of tourist attractions.

A further goal was reached in 1987 with the completion of the mapping of the whole Soviet Union at the scale of 1:25,000 (about 200,000 sheets).

THE GLOBAL PROJECT

But the mapping of the USSR is only a small part of the story. It is difficult to grasp the immensity of the Soviet military global mapping project: the VTU conducted a secret topographic mapping program at a high level of detail and coverage for almost the entire globe. The true extent of the Soviet cartographic enterprise has yet to emerge, but it is clear that this was the most comprehensive global topographic mapping project ever undertaken. The number of different maps produced is impossible to quantify, but one estimate [50] puts the figure at well in excess of one million.

The map series can be classified as follows:

- Topographic maps (topos):
    - Military series (SK-42)
    - Civil series (SK-63)
- City plans:
    - Military series
    - Civil series
- Special maps, such as 1:300,000 topographic maps, large-scale small-town plans, aeronavigation maps, and rectangular topographic maps

SYMBOLOGY

The map series are described more fully below. What they all share, however, is the most comprehensive system of symbology and annotation ever devised. All mapping relies on the establishment of, and adherence to, a standardized and uniform policy covering what is to be mapped, to what level of accuracy and detail, and how—through a "vocabulary" comprising hundreds of cartographic symbols—to show it.

The Soviet symbology and specifications, which evolved substantially over the period of the 1940s to the 1990s, were designed as a single all-embracing system, applicable worldwide, and adaptable for all scales and series of maps. They portray terrain and communications and identify cultural features in a standardized fashion to enable the map user to very quickly become familiarized with the landscape depicted.

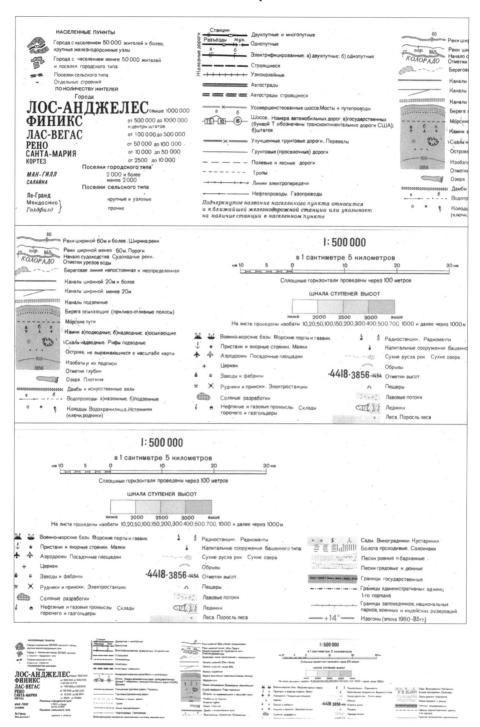

2.3 Marginalia from 1:500,000 sheet J-12-4, *Cortez*, Colorado, 1981, showing the simplified depiction of the legend appearing on unclassified small-scale map sheets.

Many hundreds of specific symbols were devised to differentiate in as much detail as possible the purpose and construction of individual buildings, the religion of places of worship, the type and density of vegetation and crops, and the nature of the terrain and coast. Appendix 5 presents a small sample of these symbols and annotations. Colors and hachuring are used to identify, for example, built-up areas where non-fireproof buildings predominate, or city blocks where the majority of buildings are high-rise multi-story structures. A hierarchy of about twenty classes of size and style of lettering of names is used to denote the size and status of towns and cities; similarly, navigable rivers are named in uppercase letters, non-navigable in lowercase.

Of particular interest to the cartographer and the potential map user were means of travel and "the going" (the ease of traversing the terrain). Railroads, roads, mountain passes, ferries, and bridges are indicated with as much detail as possible, as are forests and rivers that might impede progress. For this purpose, the distinct symbols and colors are supplemented by a convention of annotation, whereby important dimensions, characteristics, and numeric values are shown alongside.

The depiction of railroads is enhanced with information such as the number of tracks and whether or not electrified, and the position and importance of station buildings. For roads and tracks, the quality, number, and width of carriageways and surface material are shown together with the overall width of clearance. Similarly, the months when mountain passes are open and the dimensions and carrying capacity of ferries and bridges are annotated, as are the type of trees and their typical height, girth, and the clearance between them in forests, and also the speed of flow, depth, and bed of rivers. Of course, not all the specified information could be collected in every case; what is surprising is how much of this hard-to-obtain detail is shown on maps of non-USSR territory. This gathering of data is discussed in chapter 3.

The symbols, the colors, and the annotation conventions are minimally depicted in the marginalia of smaller-scale sheets (fig. 2.3) and generally not at all on the large-scale topos, but are defined in officers' guides [39] and cartographers' handbooks [40], the latter amounting to some 220 pages. A series of simplified colorful wall posters was produced for training purposes (figs 2.4 to 2.8). Some versions of the tables of symbols have been translated into English [1, 7, 9].

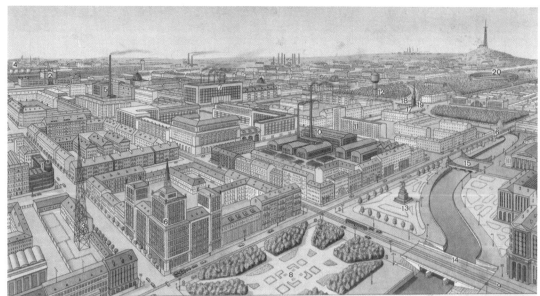

2.4 Detail from training poster 1, "Communities," 1968. The posters are all approximately 900 mm × 580 mm.

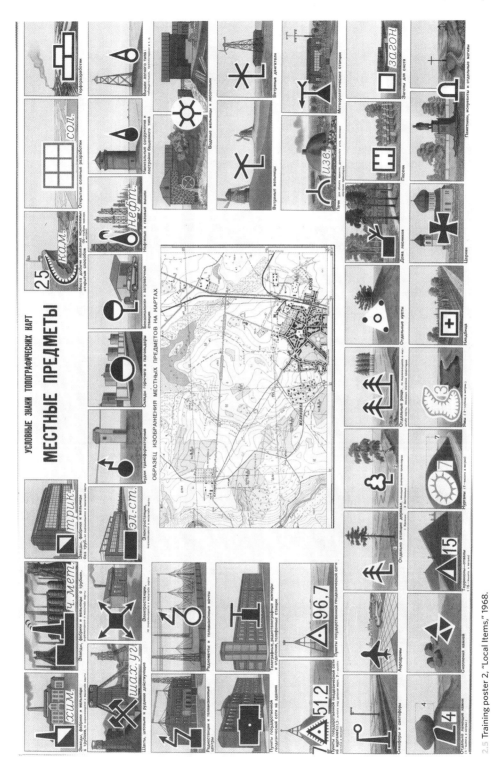

2.5 Training poster 2, "Local Items," 1968.

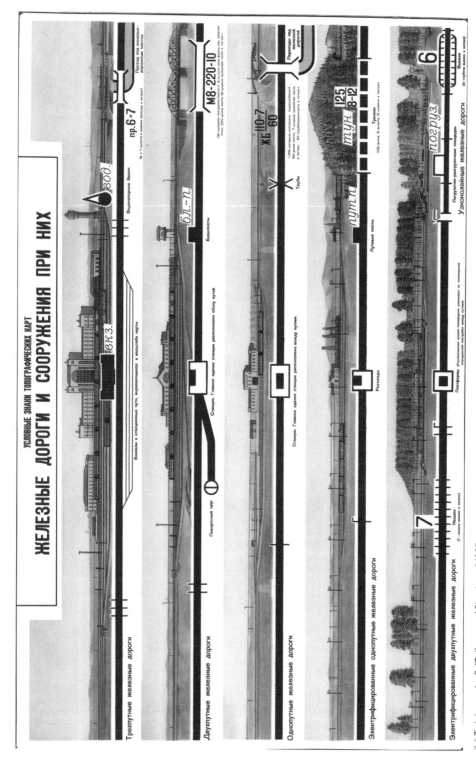

2.6 Training poster 3, "Railways and Structures," 1985.

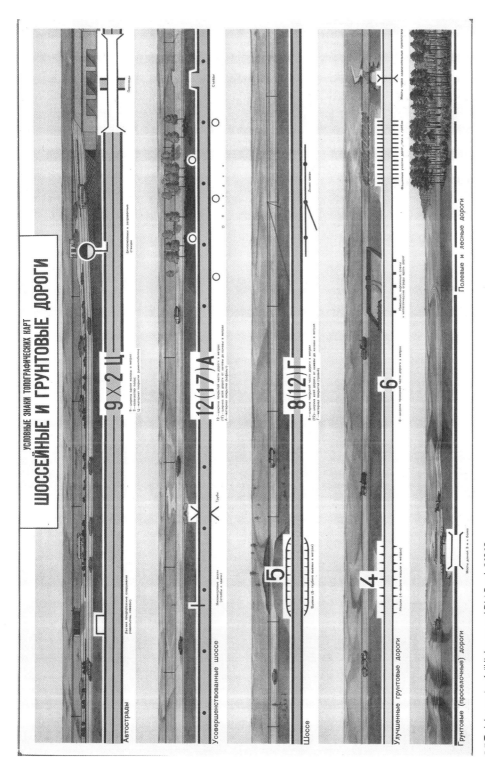

2.7 Training poster 4, "Highways and Dirt Roads," 1968.

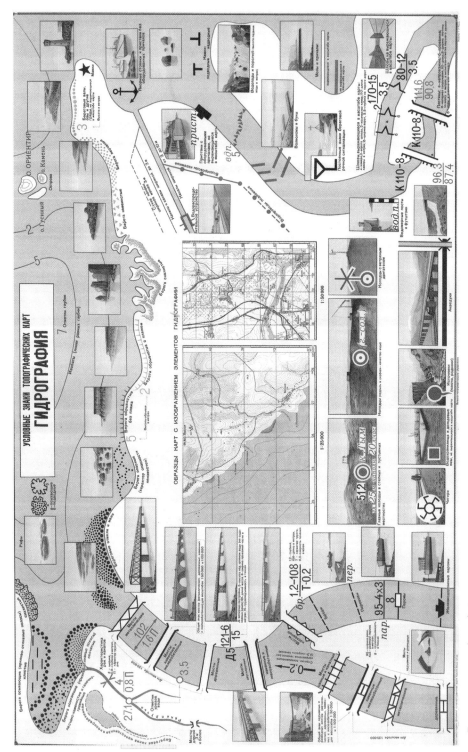

2.8 Training poster 5, "Hydrography," 1968.

THE MAP SERIES

SK-42 Topos

The topographic maps, or "topos," are non-rectangular sheets based on sheet lines defined by lines of latitude and longitude and numbered in accordance with the nomenclature devised for the International Map of the World (IMW). Topos were produced at seven scales, classified by the Soviet authorities as follows:

- Small-scale/general terrain evaluation: 1:1,000,000
- Small-scale/operational: 1:500,000
- Medium-scale/operational-tactical: 1:200,000
- Medium-scale/tactical: 1:100,000
- Large-scale/tactical: 1:50,000, 1:25,000, 1:10,000

The security classifications of the topographic maps are in general as follows:

- Small-scale sheets are unclassified.
- 1:200,000-scale maps are labeled "Для Служебного Пользования" (For Official Use).
- 1:100,000- and 1:50,000-scale maps of USSR territory are labeled "Секретно" (Secret).
- Those of the rest of the world are labeled "Для Служебного Пользования" (For Official Use).
- Larger-scale topos and all city plans are labeled "Секретно" (Secret).

However, this general rule is not absolute and exceptions are found. Figures 2.9A and B show the classification found on typical sheets.

The basic quadrangle is the 1:1,000,000 sheet spanning 4° latitude by 6° longitude. The quadrangles are identified by lettered bands north from the equator and by numbered zones east from longitude 180° (see figs. 2.10A and B). Thus London, on sheet M-30, lies in band M (the 13th band north, latitudes 52° to 56°) and in zone 30 (0° longitude, the Greenwich meridian being the boundary between zones 30 and 31). South of the equator the same system applies, with the letters preceded by "s." The bands use letters of the Roman rather than the Cyrillic alphabet, although Cyrillic is used for the subdivisions into larger scales, described below.

As lines of longitude converge toward the poles in the projection used, so the width of the northerly topos grows narrower, such that in band P,

2.9A "СЕКРЕТНО" (Secret) classification at top right of 1:50,000 sheet O-35-074-3 of USSR territory (Matishi, Latvia), 1990.

2.9B "ДЛЯ СЛУЖЕБНОГО ПОЛЬЗОВАНИЯ" (For Official Use) classification at top right of 1:50,000 sheet M-30-041-3, *Okehampton*, UK, 1980.

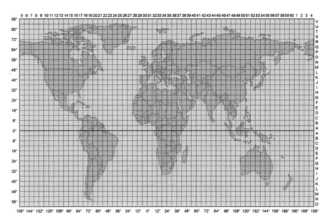

2.10A 1982 index map showing 1:1,000,000 quadrangle system.

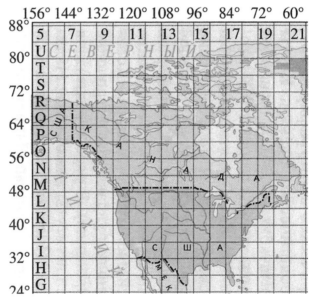

2.10B Detail of 1:1,000,000 grid for North America.

north of latitude 60°, two maps are printed as a single sheet. Sheet P-29/30 is an example of such a double-width sheet, covering the Faeroe and Shetland Islands. It is about 700 millimeters wide, similar to single sheets in band H.

Each 1:1,000,000 sheet is subdivided into four 1:500,000 sheets (from northwest to southeast), labeled А, Б, В, and Г, respectively (the first four letters of the Russian alphabet). K-18, for example, stretches from latitude 40° to 44° N and longitude 72° to 78° W. New York, at approximately 40°40′ N and 74°00′ W, lies in the southeastern quarter of K-18 and hence on 1:500,000 sheet K-18-Г.

The map sheets themselves are titled "Генеральный Штаб" (General Staff), with a map name and two versions of the reference number—the "original" in black (e.g., K-18-Г) and a second all-numeric version in blue (in this case, 11-18-4), which eliminates possible ambiguity (see figs. 2.11A and B). The convention that will be used in this book is to retain the band letter and use numerics for the rest of the code—namely, the zone number (1–60) followed by

- one digit for 1:500,000 sheets (1–4);
- two digits for 1:200,000 sheets (01–36, see below);
- three digits for 1:100,000 (001–144, see below).

2.11A "ГЕНЕРАЛЬНЫЙ ШТАБ" (General Staff) and sheet name, *New York*, on 1:500,000 sheet K-18-4, 1981.

2.11B Sheet K-18-4, shown as K-18-Г and 11-18-4 (single digit in third position indicating the scale of 1:500,000).

This convention avoids confusion for non-Russian speakers caused by Cyrillic Б/б (the second letter of the alphabet, which is Roman B/b) and Cyrillic В/в (the third letter, Roman V/v).

Each 1:1,000,000 sheet is also subdivided into 36 1:200,000 sheets in a six-by-six grid that is numbered in Roman numerals from I to XXXVI. The city of Riga, Latvia, for example, is on sheet O-35-XXV, referred to here as sheet O-35-25 or, on the map itself, 15-35-25.

The 1:200,000 sheets normally contain on the reverse side a detailed written description of the district (towns, communications, topography, geology, hydrology, vegetation, and climate) together with a geological sketch map. See figure 2.12 for the reverse of the Riga sheet and appendix 4 for a translation of an example from sheet N-31-31, Cambridge, UK. Examples from Afghanistan, Lebanon, Syria, and Ukraine have been published as compilations by East View Press [12, 13, 14].

The successive subdivision into sheets at larger scales continues. There are 144 1:100,000 maps to each 1:1,000,000 sheet, in a twelve-by-twelve grid numbered 1–144 (or 001–144), as in figure 2.13A. Each of these 1:100,000 sheets is subdivided into four 1:50,000 sheets (А, Б, В, Г), and, where applicable, each of these into four 1:25,000 sheets (lowercase а, б, в, г), as in figure 2.13B.

2.12 Reverse side of 1:200,000 O-35-XXV (O-35-25), *Riga*, Latvia, showing the information typically given on reverse of maps this scale: "СПРАВКА О МЕСТНОСТИ" (Information about the Area) and a geological diagram.

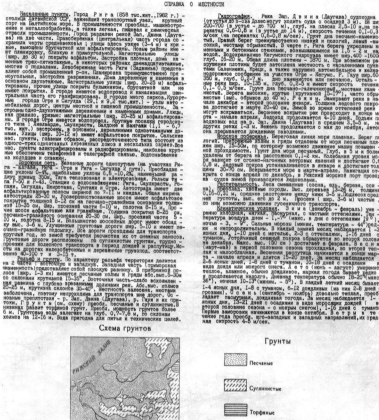

These each subdivide further into four 1:10,000 sheets (1, 2, 3, 4). The two largest scales were only produced for USSR territory.

The system immediately locates any sheet precisely on the surface of the globe. For example, 1:10,000 sheet O-35-87-Г-а-1 (or O-35-087-4-1-1), shown in figure A1.55, can readily be identified as having its northwest corner at 57°30′ N and 25°15′ E. At latitude 54°, the approximate size of area covered by each sheet is as follows:

1:1,000,000: 175,000 km^2 1:100,000: 1,200 km^2
1:500,000: 44,000 km^2 1:50,000: 300 km^2
1:200,000: 5,000 km^2 1:25,000: 75 km^2

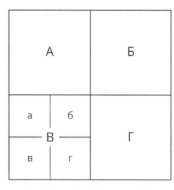

2.13B Breakdown of each 1:100,000 sheet into four 1:50,000 sheets and each of these into four 1:25,000 sheets.

2.13A (*Left*) Index diagram showing breakdown of each 1:1,000,000 quadrangle into 4 × 4 grid of 1:500,000 sheets, 6 × 6 grid of 1:200,000 sheets, and 12 × 12 grid of 1:100,000 sheets.

Although some topos were produced before the Second World War (e.g., sheet M-30 of 1938, shown in fig. A1.38), the specifications evolved and were gradually enhanced. Most of the global mapping project was undertaken during the period of the 1950s to the 1990s. Many have successive editions produced at occasional intervals, typically ten to twenty years, as shown in figures 2.14A and B.

Maps were also reprinted where required, with no changes to the content but with new print codes, as in figures 2.15A, B, and C.

The projection used in most Soviet mapping (at scales up to 1:1,000,000) is the Gauss-Krüger (G-K) conformal transverse cylindrical projection (also known simply as Gauss). Before 1984, maps at 1:1,000,000 were based on a modified polyconic projection used by the International Map of the World.

Essentially, as a conformal projection, angles on the Earth's surface are preserved, allowing bearings to be taken from the map. The graticules (lines of latitude and longitude) use Coordinate System 1942 (Система Координат 1942) or SK-42, from an ellipsoid (a measure of the Earth's figure) developed by the geodesist Feodosy Krasovsky. The origin of the

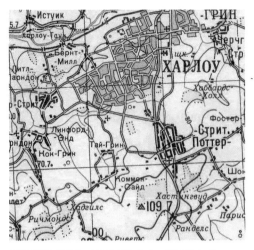

2.14A 1:100,000 sheet M-31-001, *Harlow*, UK, 1964 edition.

2.14B 1982 edition of M-31-001, *Harlow*, UK, showing growth of the town and arrival of a motorway. Note that the district of Potter Street, Поттер-Стрит, which was misnamed "Стрит-Поттер" on the 1964 sheet, is corrected.

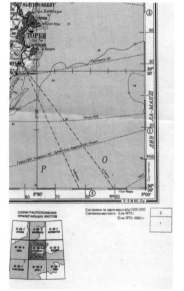

2.15C Both printings have the same edition date of 1985 in the top right corner.

2.15A Original March 1985 print code "E-3 III 85-Ср" on 1:500,000 sheet M-30-1, *Cardiff*, UK. (Incidentally, Cardiff is a strange choice for the name of this map—the city is right on the edge of the sheet. Exeter, Plymouth, and Swansea all appear, and any one of these would have been more appropriate.)

2.15B M-30-1, reprint of January 1989, with print code "E-24 I-89 Л."

coordinate system is the Pulkovo Observatory, near St. Petersburg, and elevations above sea level are referenced to the zero datum of Kronstadt on the Baltic Sea (the main port of St. Petersburg and the traditional seat of the Russian admiralty).

Gauss-Krüger is based on a regular system of Transverse Mercator projections with each covering a zone 6° wide, with central meridians (axial lines of longitude) at 3° intervals. The advantage of this is that it simplifies the depiction of the globe as a flat surface for relatively small areas and allows the use of a rectangular grid within each zone. The arrangement, however, negates the ability to join up the grids of adjacent zones; since the central meridians of each grid are not parallel to one another, an angle occurs between the two neighboring grids. Because this angle would make the correct calculation of distances between two points on neighboring plans impossible, the topographic maps that cover up to 2° east and west from the central meridian (e.g., 3° W for the zone M-30) include an additional grid for the eastern (or western) zone (i.e., M-29 and M-31, respectively), which is shown using thinner lines on the outer edges of the map border.

The maps themselves are generally printed in eight, ten, or twelve colors and usually show latitude and longitude graticules at frequent intervals (depending on scale) as ticks along the neat line (the map edge). They also have a secondary set of graticules showing G-K coordinates. The maps also contain a grid; for small-scale sheets, this is based on latitude and longitude lines (fig. 2.16A); for medium- and large-scale maps, the grid is based on G-K values and is not parallel to the sheet edges (fig. 2.16B).

Most topos include two key diagrams: one locating the sheet and its neighbors, the other delineating civil administrations (countries, states, counties, and so on) appearing on the map (see fig. 2.17). All sheets have a print code showing the factory and date of printing, and most sheets also show compilation and revision dates, sometimes with cartographers' and editors' names. See appendix 7 for details of print codes and figures 3.1, 3.2, and 3.3 for examples of compilation data.

Other Warsaw Pact countries published topos of their own territory (e.g., see figure 4.5), and Czechoslovakia, East Germany, Hungary, and Poland, at least, also produced maps of parts of Western Europe (figs. 2.18A, B, and C).

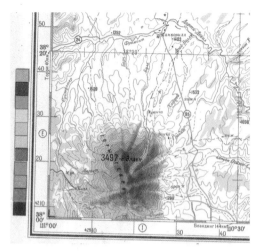
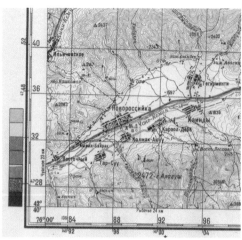

2.16A Lower left corner of small-scale (1:500,000) sheet J-12-2, *Grand Junction*, Colorado, showing color registration block, latitude and longitude values and grid, and Gauss-Krüger graticules.

2.16B Lower left corner of medium-scale (1:200,000) sheet K-43-11, *Almaty*, Kazakhstan, showing latitude and longitude values, two sets of Gauss-Krüger values (the outer set being applicable to the adjacent zone), and a G-K-based grid.

2.17 Lower right of 1:500,000 sheet K-17-2, *Toronto*, 1981, showing diagram of political boundaries, with Toronto identified as provincial center (o). Political entities are listed as follows: Canada: 1. Province of Ontario; USA: 2. Ohio, 3. Pennsylvania, and 4. New York. The diagram of neighboring sheets shows L-17-В, *Alpena*, Michigan; L-17-Г, *Peterborough*, Canada; L-18-В, *Ottawa*, Canada; К-17-А, *Detroit*, Michigan; К-18-А, *Rochester*, New York; К-17-В, *Cleveland*, Ohio; К-17-Г, *Pittsburg*, Pennsylvania; and К-18-В, *Scranton*, New York.

SZTAB GENERALNY W.P.　　　　　　　　　　　　　　　　　　　　　TAJNE
　　　　　　　　　　　　　　　　　　　　　　　　SOUTHAMPTON M-30-XI

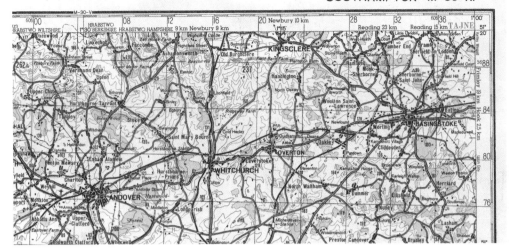

2.18A 1:200,000 sheet M-30-11, *Southampton*, UK, 1967, published by Polish General Staff and printed by WZKart. The sheet is classified "Tajne" (Secret). Source data is stated as 1:63,360 map of 1947 (which is the openly available Ordnance Survey 1-inch-to-the-mile mapping) and 1:200,000 map of 1957 (possibly a Russian edition).

2.18B Polish version of 1:1,000,000 sheet M-31, *Paris*, 1957, publisher unstated and with no security classification, printed by WZKart. The sheet is headed "NAZWY W PISOWNI FONETYCZNEJ" (Names in Phonetic Spelling). Unlike figure 2.18A, the British placenames are rendered phonetically for a Polish speaker (ISTBON is Eastbourne, HEJSTYNZ is Hastings, MAGYT is Margate, and LANDEN is London).

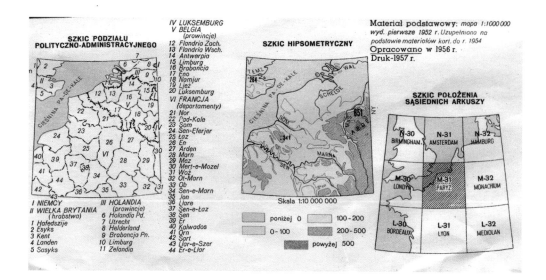

2.18C The lower right of Polish M-31. The diagram of neighboring sheets has M-30 as "Londyn" (the Polish name for London), rather than "Landen," as it appears on the map itself. Note the hypsometric diagram and the delineation of British counties, Dutch and Belgian provinces, and French départements.

SK-63 Topos

The USSR and its allies had need of topographic mapping for economic and civil governmental purposes, but the military maps, having the location-based sheet numbering system and showing geographic coordinates—designed to be used for military-precision targeting—were not suited to being made available to non-military authorities. This problem was resolved by the introduction of the Coordinate System 1963, SK-63. For this purpose, USSR territory was arbitrarily divided into twenty regions, each with its own code letter. Table 2.1 shows the 1963 series region codes.

Fig 2.19A shows the allocation of SK-63 codes across the USSR, apparently randomly distributed between latitude bands J to V and longitude zones 34 to 62. The breakdown into 1:100,000 sheet lines for Latvia is shown in figure 2.19B.

SK-63 is an undistorted derivative of SK-42, but sheet lines are inconsistent with SK-42; the sheets are numbered with the "region" codes rather than the IMW codes and have no other title and no coordinates, just an unlabeled grid. They would be useless to anyone who did not already know the identity of the location mapped or who needed any precise reference point. These maps were produced by the GUGK rather than by the VTU.

In other respects, the maps are similar to the military editions, having the same symbology and identical detail and being classified as "Secret." Only sheets of scales 1:100,000, 1:25,000, and 1:10,000 were published, and

TABLE 2.1

A	Georgia	Q	Northwest
B	Almaty Oblast	R	Central and Lower Volga (Kalmykia, Astrakhan, Volgograd, Saratov, Penza, and Ulyanovsk Oblasts)
C	Baltic States and Northwest		
D	Udmertia and Perm Oblasts		
E	Altai	S	Yakutia and Magadan Oblasts
F	Irkutsk Oblast	T	Krasnodar Krai
G	Khabarovsk Krai, Sakhalin Oblast	U	Kirghizia, Tadzhikistan, Turkmenia, Uzbekistan
I	Novosibirsk, Omsk, and Tomsk Oblasts		
J	Kamchatka	V	Bashkiria, Tatarstan, Samara, Orenburg, Chelyabinsk, Kurgan Oblasts
K	Kazakhstan		
L	Tiva and Krasnoyarsk Krai	W	Sverdlovsk and Tyumen Oblasts
M	Altai Krai, Kemerovsk Oblast	X	Ukraine and Moldavia
P	Central Black-Earth Region of Russia	Y	Kazakhstan

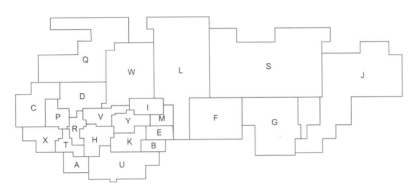

2.19A Diagram of the allocation of SK-63 zone codes across the USSR.

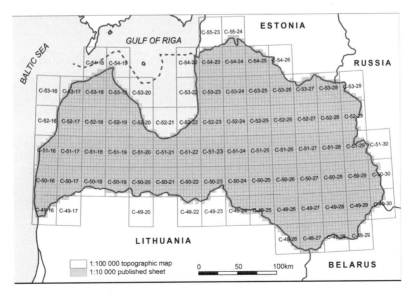

2.19B Sheet lines of 1:100,000 SK-63 sheets of Latvia in zone C.

CAPTURING THE WORLD—ON PAPER

W-27-31-B-a СЕКРЕТНО

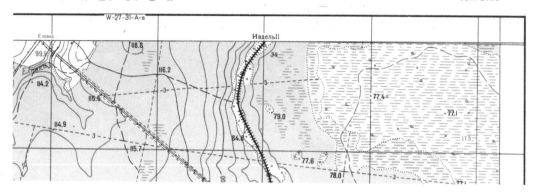

2.20 Top edge of SK-63 1:25,000 sheet W-27-31-B-a (W-27-031-3-1), 1980, published by the GUGK. This covers an area north of the town of Serov in Sverdlovsk Oblast on SK-42 sheet O-41-014-1.

only of USSR territory. Examples are shown in figures A1.47, A1.54, and A1.56. Another example is shown in figure 2.20, labeled W-27-31-B-a "Секретно" (Secret), with no map title or other geographic identifier.

In East Germany, the equivalent to SK-63, maps for civil authorities with reduced map content and positional inaccuracies, were known as AV (Ausgabe für die Volkswirtschaft, or Edition for the National Economy). They are described in the volume edited by Dagmar Unverhau [32]. Other Warsaw Pact countries established similar local coordinate systems; the JTSK in Czechoslovakia, the EOTR in Hungary [26], the 1965 system and the GUKiK-80 in Poland [30], and the Stereo-70 in Romania.

CITY PLANS—MILITARY SERIES

The city plan series differs from the regular topographic map series in three important respects:

    1. The sheets are rectangular, the edges being defined on Gauss-Krüger coordinates.

    2. The coverage is non-continuous across a country, as the plan of each chosen city is designed as a single entity so that a set of sheets is centered on and covers just the required urban area.

    3. They generally have a street index, a written description of the locality (Справка, or spravka), and "important objects" highlighted and listed.

City plans use the G-K projection and Coordinate System 1942. It is

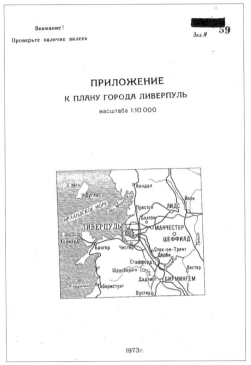
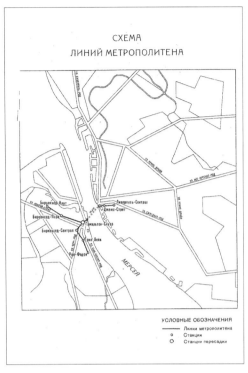

2.21A Cover of supplement booklet with index (spravka) for *Liverpool*, UK, 1:10,000 city plan, 1973.

2.21B Diagram of metro lines in Liverpool booklet (the Mersey Railway).

impossible to know definitively which cities were mapped (or even how many)—we only know of those that have surfaced in the West in recent years. Those known include about two thousand cities worldwide (not counting any in Russia itself), of which about 120 are in the United States and 100 in the United Kingdom. There are sure to have been many more, which, like those of Russia, have not found their way into Western awareness. The earliest known sheets date from the 1940s, while the great majority of them were produced in the period 1960 to 1990. Lists of known plans can be found online [51].

Most are either 1:25,000 scale (usually large conurbations) or 1:10,000 scale, but a few examples are known of scales 1:5,000, 1:15,000, and 1:20,000. In some cases—such as London, Liverpool, New York, and others—the street index, spravka, and list of important objects were published as a separate booklet (figs. 2.21A and B).

2.22 The footprints of the known city plans covering the British Isles.

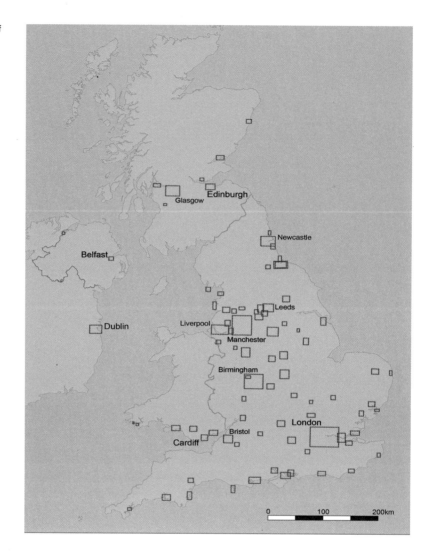

The list of cities known to have been mapped includes several relatively small, economically unimportant places as well as the obvious major centers. This fact suggests that there were many more city plans produced than are currently known. In the British Isles, for example, a plan exists of the rural market town of Gainsborough, Lincolnshire, whereas none has been found of more significant places such as Carlisle, Kingston upon Hull, or, in the

Republic of Ireland, the major cities of Cork and Limerick. These and others may well have been produced and may yet emerge.

The relatively close proximity of urban areas in Britain gives rise to overlaps, in which areas between cities are included in more than one plan (see fig. 2.22). This is explored in more detail in chapter 3, where it can be seen from the discrepancies and differences that the compilers of each plan were working independently and not making use of common data. Similarly, new editions were published from time to time, and the evidence examined in chapter 3 shows that these were generally new productions, not simply revisions of existing sheets. In some cases, the new plan is at a different scale (1974 Bournemouth, UK, is on four sheets at 1:10,000, while the 1990 edition is one sheet at 1:25,000) or has different boundaries (1973 Luton, UK, is one 1:10,000 sheet, while the 1986 edition extends to a larger area on two sheets).

There is no standard sheet size (and, of course, no standard area of coverage). The composition is central to the city—the area to be mapped is arbitrarily divided into several suitably sized sheets. These vary from about 500 to 1,500 millimeters in each direction, with the amount of blank space outside the sheet margins varying from about 50 to 150 millimeters. The maps may appear in landscape or portrait format on the sheet, and the sizes of the sheets comprising a single city set may be different.

Typically, a city may require two or four sheets, with the division between them falling in the city center, which may have been inconvenient for use in the field. The largest number of sheets for any conurbation seems to be the twelve-sheet set of Los Angeles. New York and San Francisco each have eight sheets; Chicago has seven (these US examples are at 1:25,000 scale). As the component sheets are rectangular, the margins could have been trimmed off and the sheets mounted side by side to produce a single composite wall map, if required (unlike the topo maps), as shown in figure 2.23.

The city plans generally have a Gauss-Krüger grid, to which the sheet edges are aligned, showing the G-K coordinate values, together with latitude and longitude values shown as ticks along the margins. However, plans of East German and Polish towns generally have the sheet edges aligned to lines of latitude and longitude, with G-K-based grid lines running non-parallel to the edges. Cities close to G-K zone boundaries have a second set of G-K

ГЕНЕРАЛЬНЫЙ ШТАБ
ЛОНДОН

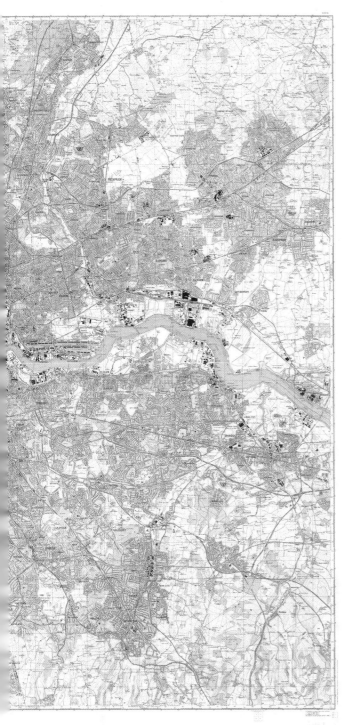

2.23 A composite of the four sheets of the city plan of *London*, 1:25,000, 1982, with the inner margins removed to show the unified composition.

2.24 Corner of *Darlington*, UK, 1:10,000, 1976, showing second set of Gauss-Krüger coordinates (applicable to the adjacent G-K zone) outside the margin.

2.25 Title of *Beijing*, 1:25,000, 1987, showing the publisher (General Staff), the city (Peking), and the 1:100,000 topos applicable. Note that the edition date is 1983 although the print date is 1987.

ГЕНЕРАЛЬНЫЙ ШТАБ
ПЕКИН
(К–50–137,138; J–50–5,6)
Издание 1983 г.

coordinates shown along the map edge, for ease of use with maps from an adjacent zone (see fig. 2.24).

Deriving coordinates from the grid shown on the city plans is not as straightforward as it might seem. Northings are given by the number of kilometers from the equator (e.g., 6050 in the northeast corner of the plan of Darlington in fig. 2.24). The last three digits of the eastings refer to the number of kilometers east from the false meridian (to avoid negative coordinates), which is located 500 kilometers east from the central meridian of the zone (i.e., 3° W for zone N-30). The remaining digits refer to the number of the zone minus 30, with the numbers restarting at 60 after zone 30 (i.e., 0) is reached. The number 60598 in the same corner means that this point is located in zone 30 (30 – 30 = 60), at a distance of 598 – 500 = 98 kilometers to the east from the central meridian, which is 3° W for this zone.

City plans have the city name placed centrally across the entire (multi-sheet) map together with General Staff, the reference number of the 1:100,000 SK-42 topographic sheet(s) on which the city appears, and the edition date, as in figure 2.25.

Contour lines may be as close as 1 or 2 meters in USSR territory, or more usually elsewhere at 2.5-, 5-, or 10-meter intervals; in a few cases, they are at 30 meters (e.g., Dublin, Ireland).

The specification for the number of colors used and the detail depicted evolved over time as printing technology and data-gathering techniques improved. The earlier plans (1940s–1960s) are printed in four colors and have relatively sparse detail (examples are seen at figs. 1.1, 3.4, 3.8, A1.6, A1.36).

Later plans (those from the 1970s onward, as most of the other extracts in this book) are printed in eight, ten, or twelve colors and include as much of the detail required by the standard specification as has been possible

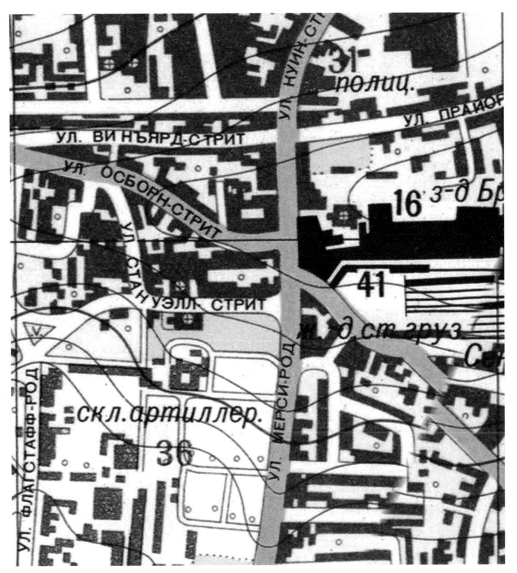

2.26A Extract from *Colchester*, UK, 1:10,000, 1975, showing administration object 31, highlighted in purple and listed as "Police"; industrial objects 16 and 41, highlighted in black and listed as "factory—unknown production" and "railway freight station St. Botolph," respectively; and military object 36, highlighted in green and listed as "artillery depot."

2.26B Lower left of *Colchester*, UK, plan depicting the three colors with a key. Purple: government and administrative offices and their numbers; green: military facilities and communication facilities and their numbers; black: military-industrial facilities and their numbers.

to collect. They show street names, metro stations, tramlines (streetcars), and either the footprint of individual buildings or the generalized shape of each city block, differentiating where possible between high- and low-rise blocks. The extent and accuracy of such detail is discussed in chapter 3, together with suggestions as to how the data could have been gathered.

Specific "important objects" are individually identified, highlighted in color, numbered, and listed in the index. Industrial buildings such as factories, railroad stations, public utilities, and so on are shown in black (for factories, the company name and its products, if known, are also included in the index). Facilities for civil administration are colored purple, and objects of military significance are shown in green (see figs. 2.26A and B).

Considerable effort has evidently been applied in acquiring and identifying these important objects—500 are listed for Los Angeles, 395 in Birmingham, UK, 374 in London, and 314 in Boston (see table 3.2). Objects are listed alphabetically and then numbered sequentially, which means that all must be identified before any can be numbered and marked on the map.

The list of important objects usually appears on one of the plan sheets (or in a separate booklet), alongside the street index and the spravka—a descriptive essay of one to two thousand words describing the city and its environs under headings such as "General Information," "Industry and Transport," "Economy," "Topography," "Climate," and "Communications." A translation of a typical city spravka, that of Cambridge, UK, appears in appendix 3.

The marginalia of city plans generally includes some compilation information and always has a print code identifying the date and factory of printing (print codes are deciphered in appendix 7).

CITY PLANS—CIVIL SERIES

The GUGK produced a "parallel" series of city plans for use by civil government in the USSR, with "Local Coordinate Systems." These are to the same specification, coloring, and symbology as the military city plans, but omit geographic coordinates, the street index, spravka, and important objects.

While the local coordinates were established separately (and secretly) for each map, examination of the plan of Valmiera, Latvia, and comparison with

TABLE 2.2

Version	Military	Civil
Published by	General Staff	GUGK
Title	Valmiera O-35-87	Valmiera
Date	1975	1990
Security classification	Secret	Secret
Coordinate system	SK-42	local coordinate system
Ground coverage	8 × 8 km	8 × 8 km
Contour interval	2.5 m	2 m
Latitude/longitude coordinates	shown by ticks on margin at 2′ intervals	not shown
G-K coordinates	values shown for two zones	not shown
Grid	16 × 16 grid parallel to edges, numbered 1 to 16 west to east and lettered A to P (in Cyrillic alphabet) north to south	approx. 8 × 8 grid not parallel to edges, numbered 1 to 8 west to east and 8 to −1 north to south (the line labeled zero runs diagonally east to west near the southern edge of the plan)
Roads and railroads crossing the map edge	destination and distance shown	destination only
Annotation of roads, forests, rivers, spot heights, etc.	very detailed	very detailed
Street index and spravka	both	neither
Important objects highlighted and listed	no	no

the standard military plan shows that the projection is the same as the SK-42 projection, but that identifying values are omitted and a local grid applied. The two Valmiera plans, both at 1:10,000 scale, cover the same ground (the four corners being in the same locations) and are very similar in general appearance and content (see figs. 2.27A and B). There are, however, several differences, as table 2.2 shows.

The 1:10,000 Local Coordinate System plans published by the GUGK of the Latvian cities of Ventspils (1978) and Jēkabpils (1979) are similar to the Valmiera example, except that the grid is parallel to the edges, the contour interval is 2.5 meters, and distances are given where roads cross the map edge. The 1:10,000 Riga six-sheet set of 1988 is as described for Valmiera, but only black-and-white, rather than color, copies are known. All are classified as "Secret."

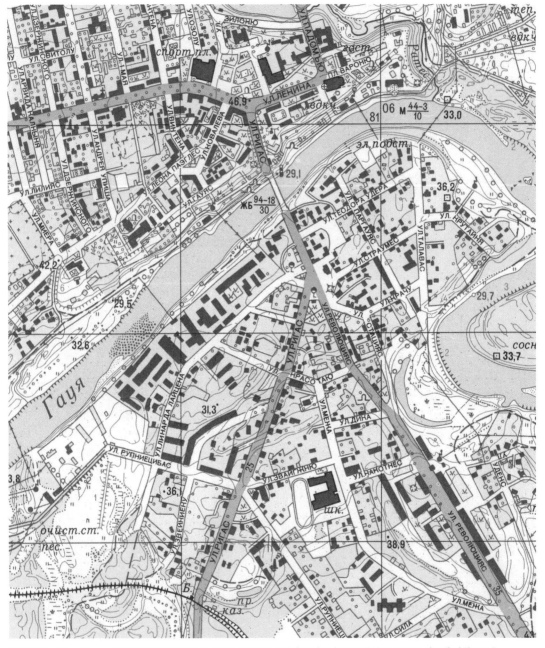

2.27A Extract from General Staff city plan of *Valmiera*, Latvia, 1:10,000, 1975, based on System 1942 projection, classified "Secret."

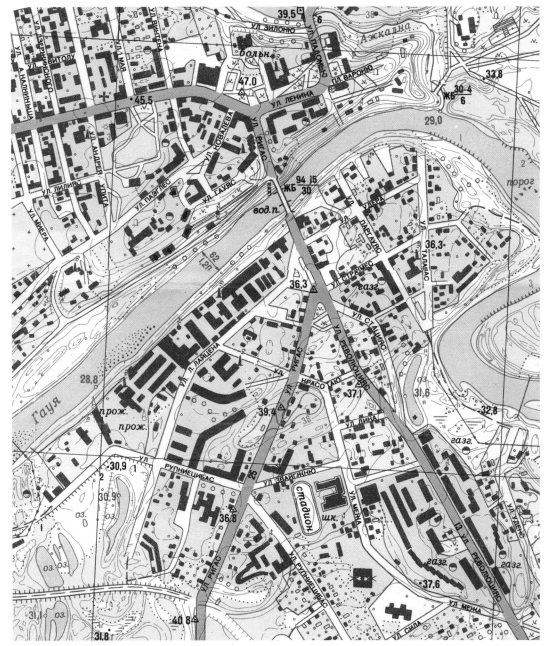

2.27B Extract from GUGK city plan of *Valmiera*, Latvia, 1:10,000, 1990, based on "Local Coordinate System," classified "Secret," and intended for use by civil authorities. The general appearance, level of detail, and projection are similar to the General Staff version but updated to the latest conditions.

SPECIAL MAPS

Other map series are known to have been produced by the VTU and the GUGK during the Cold War period. They include 1:300,000 topos, such as VI-K-39 of 1949, showing the Caspian Sea and the Uzbekistan-Turkmenistan border (fig. 2.28). The code VI-K-39 indicates that this is sheet six of the nine 1:300,000 sheets covering quadrangle K-39. The purpose of this civil series, published in the 1950s by the GUGK, is unknown, nor is it known how extensive the series was; but the few that have come to light cover areas within the USSR, ranging from J-42 Tajikistan, in the south, to P-40 Krasnovishersk, Russia, in the north, and the Baltic states. In the section describing 1:200,000-scale maps, the 1958 USA Army Technical Manual [9] includes the statement: "the symbolization is [. . .] believed to also apply to the new 1:300,000 scale maps."

Also of interest are the series of large-scale town plans labeled "План-Схема" (Plan-Scheme) and classified "Для Служебного Пользования" (For Official Use), produced by the GUGK. About thirty-five plans are known of small Latvian towns at scales of 1:5,000, 1:6,000, 1:7,000, 1:8,000, 1:10,000, 1:18,000, 1:20,000, and one of Riga at 1:25,000, all dating from the mid-1980s. They show little detail, being a much simplified design in four colors, with no detailed symbology or annotation (see fig. 2.29). Few individual buildings are shown, urban blocks are shown generalized, and contours are at 5- or 10-meter intervals, with no spot heights or depths. They do have a street index but no spravka. They also show the symbol legend on the face of the map, facilitating their use by untrained users.

The Riga sheet, dated 1984, has an enlarged inset showing the city center (scale not stated), and district names are printed in red, delineated with red broken lines. A variation is the 1:20,000 plan of Almaty, Kazakhstan, of 1994, which differs in that it is oriented east to west and that each district of the city is shaded a different color.

TWO OTHER MAP SERIES ARE SMALL-SCALE AERONAVIGATION MAPS and rectangular topographic maps. The aeronavigation series has large rectangular sheets at the relatively small scale of 1:2,000,000, each covering a considerable area and using a polyconic projection. Unlike the regular topos, the sheet numbers use the Cyrillic alphabet and Roman numerals.

The numbering system adopts twelve horizontal bands from north to south (A in the Arctic to M in the Antarctic) and twenty vertical zones (from zone I at London, running east via the USSR and the United States, to zone XX west of Ireland).

Sheet Б-II, Berlin, extends some 1,500 kilometers from north of Trondheim, Norway, to south of Dortmund, Germany. The next sheet to the east is Б-III, Moscow (see fig. A1.57). As they are based on a conical projection, sheets are oriented such that the central meridian (for Б-II, longitude 15°) is vertical, and lines of longitude and latitude taper north to south and curve east to west, respectively (see fig. 2.30). On sheet Б-III, longitude 33° is the central meridian.

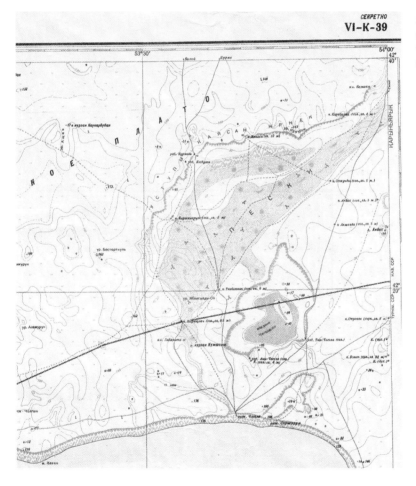

2.28 Part of 1:300,000 sheet VI-K-39 dated 1949, published by the GUGK.

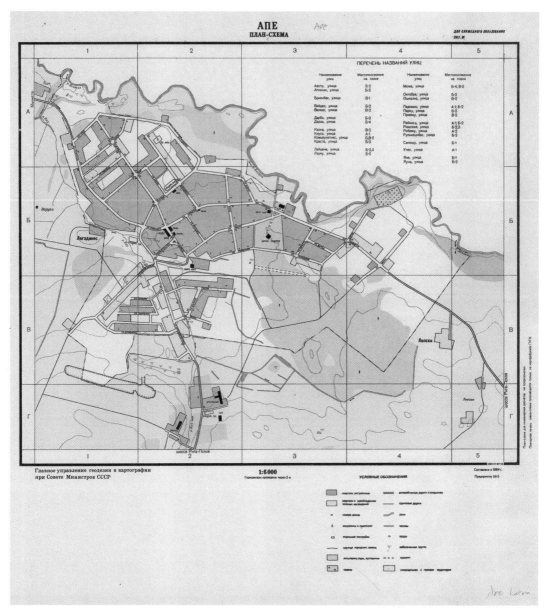

2.29 1:6,000 "plan-scheme" of *Ape*, Latvia, 1987, classified "For Official Use." Note the lack of detail.

2.30 Index diagram from 1:2,000,000 aeronavigation sheet Б-II of 1983 showing sheets to the north having an A prefix and sheets to the south having a B (A, Б, and B being the first three letters of the Cyrillic alphabet). The diagram illustrates that this sheet covers IMW bands N, O, and P and zones 32, 33, and 34, plus parts of neighboring zones where lines of longitude taper. The boundary between zone 31 and 32 is at longitude 6° and between zones 34 and 35 at 24°.

2.31A Index diagram from rectangular topo 14-00-68, *London-Paris*, 1974 (extract from this map is at figure A1.58). The diagram shows the grid lines of the regular SK-42 topos running diagonally.

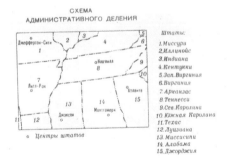

2.31B Index diagram and state boundaries and capital cities diagram from rectangular topo 44-113, *St. Louis*, Missouri, 1969.

Due to their projection, the northern edge of sheet Б-II extends from longitude 2° to 28°, while the southern edge covers only 6° to 24°. The northern edge of the Moscow sheet extends from 20° to 46°, thus providing considerable overlap, including the Baltic states and southern Finland. The extent of the southern edge of Б-III ranges from longitude 24° to 42°. These sheets carry no security classification.

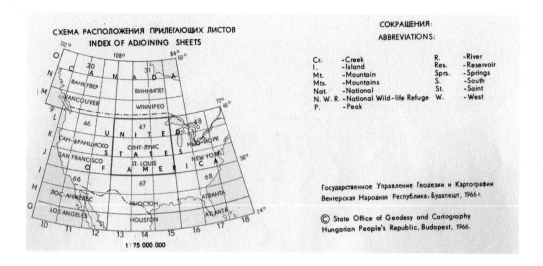

2.32 Index diagram from 1:2,500,000 World Map, sheet 47, *St. Louis, Missouri*, 1966.

Unlike the aeronavigation maps and SK-42 topos, the rectangular topos are designed to be trimmed and mounted side by side as a wall map to provide continuous coverage over very large areas (see figs. 2.31A and A1.58). These were produced in at least two different series, one in the 1960s, another starting in the 1970s, and cover at least Europe and the Americas at 1:1,000,000 and 1:500,000. The 1960s sheets are classified as "Secret," the later ones have no security classification. The 1:1,000,000-scale maps of the earlier series have two-part numbers (e.g., 44-111 is Chicago, 1969) while those of the later series have three-part numbers (e.g., 44-00-62 is Ottawa, 1985).

Although the International Map of the World (IMW), mentioned in chapter 1, was never completed, the non-secret Карта Мира–Karta Mira (World Map), at the scale of 1:2,500,000 in over 270 sheets, was produced cooperatively by the Warsaw Pact countries in the 1960s–1970s. The maps are in English and Russian and were offered on open sale in the West. Sheet 35, covering the British Isles, was produced in Berlin in 1965, while the six sheets covering the United States were produced in Budapest in 1966-68 (see fig. 2.32). Place-names are in the local language and in Roman script so, for example, all Irish places are named solely with their Irish name. Seas are named in the languages of adjoining countries; for example, "Irish Sea–Muir Meann."

PLOTS AND PLANS

ALMOST THE FIRST QUESTION TO COME TO MIND WHEN SOMEONE in the Western world encounters a Soviet military map of their hometown, having noted the amount of detail and its accuracy, is *How on earth did they do this, in such secrecy, during the Cold War?* The first obvious supposition is that the cartographers just copied existing national mapping. But there were good reasons why they did not simply do so. One reason is the natural assumption in Soviet culture that if a map is in the public domain, then it has necessarily been falsified. Within the Soviet Union, military maps of the homeland were secret and unavailable to anyone other than the military. As seen in chapter 2, even the "secret" maps for civil government were distorted, while maps for public use were little more than approximate sketch maps of tourist attractions and hiking trails. So the freely available national mapping of Western countries would be treated with caution.

Another important reason for not copying local mapping was the need to comply with a single global specification that laid down strict rules for what was to be depicted on maps. National mapping varies considerably country to country depending on national cultural expectations; huge stylistic variations are seen across Europe [21]. Soviet maps, on the other hand, had to be consistent worldwide, such that a user could pick up a map of anywhere and immediately understand the symbology, the use of colors, the naming conventions, and so on.

A Soviet cartographer tasked with compiling a map of a capitalist city, for example, would therefore make use of a whole package of materials, collected over a long period, to provide the information to show on the map.

These materials would include as much as possible of such items as official state maps, local street atlases, commercial road atlases, railway timetables, tourists' guidebooks, trade directories, aerial imagery (from spy planes or satellites), and personal reports from "boots on the ground." As can be seen from the illustrations in this chapter, some of the information derived from these sources is contradictory, some is anachronistic, some is very patchy, and some has been misunderstood or misinterpreted. Nonetheless, with the benefit of hindsight, we can see that the resulting plans were astonishingly accurate and comprehensive.

It is tempting to think of the cartographers involved in the production of these secret maps as anonymous factory workers. Yet in a few cases during the 1960s and 1970s, the names of the personnel responsible are actually recorded on the city plans and topographic maps. The extracts in figures 3.1, 3.2, and 3.3, for example, show the names of the compilers and editors. As Russian surnames indicate gender (female names end with the letter "a"), it can be seen that in the case of Belfast (fig. 3.1), the editor and both cartographers were female.

Out of several hundred sheets dating from this period examined, only 98 bore the originators' names. In total, 38 compilers are named (of which 24 were female), 42 editors are named (1 female), and 63 have the name of the unit commander or a similar title (all male and, unlike the editors and compilers, all having a military rank).

Several of the unit commanders can be seen to have been responsible for many maps over a period of several years, such as Colonel A. D. Yudin, who is named on 21 city plans (19 British cities plus Ljubljana and Maribor in the former Yugoslavia), dating from 1975 to 1980, printed in five different factories (hence, the compilation and printing were in separate locations). Colonel D. A. Mankiewicz was responsible for at least three topographic maps of the Himalayas of 1970 and for the 1976 city plan of Baden-Baden, Germany, while Colonel I. I. Shalman produced at least 8 British city plans in 1971–73.

CITING THE SOURCES

The earliest known Soviet city plans date from the 1940s. From then until the early 1960s (before the launch of Sputnik in 1957 heralded the age of the

3.1 *Belfast*, 1:10,000, printed May 1964, Leningrad. The text reads: "Plan compiled 1951 from 1:7920 plan 1947 edition and 1:10,000 plan 1940 edition, reprinted 1964. Compilers: Schischova and Rusina. Editor Galakina."

3.2 *Kilmarnock*, UK, 1:10,000, printed July 1958, Kiev. The text reads: "Compiled 1956 from map scale 10,560 edition 1910, 11 and 1:25,000 edition 1952, prepared for publication 1957, reprinted 1958. Compiler Olienikova H. A. Editor Kubyschkin A. K."

3.3 *Zurich*, 1:15,000, printed June 1952, Moscow. The text reads: "Plan compiled РКЧ 1951 from 1:15,000 plans of 1936-37 and 1:25,000 maps of 1935. Printed 1952. Compilers: Vlasov, Baryshev and Kalashnikov, Editor Schlyago."

satellite), the plans include a declaration in the margin citing the specific source mapping from which the sheet is derived. An early example is the plan of Sari, Iran, which states: "Plan compiled 1943 from 1:5000 and 1:15,000 photo-surveys of 1941 and visual reconnaissance 1943, printed 1944."

Figures 3.1, 3.2, and 3.3 show typical examples of the declaration of the sources used to derive data for the plans. In the case of British cities, the fact that the Ordnance Survey used imperial scales rather than metric scales helps to identify the sources mentioned. The Belfast plan, for example, specifies one source as "the 1947 edition of the 1:7920 plan." This is 8 inches to the mile, the same as the Ordnance Survey Belfast street plan, which was reprinted in 1947. It also cites a "1:10,000 scale plan of 1940," providing intriguing evidence of the possession by the Russians of captured German *Planheft* mapping, which was produced during the Second World War for the planned invasion of Britain. Unlike Soviet maps, the German maps were photo enlargements of standard Ordnance Survey six-inches-to-the-mile maps (1:10,560) enlarged to 1:10,000 and overprinted with additional detail, specifically targets for bombing raids (an example is seen at fig 3.64).

It is evident that these early Soviet plans were indeed derived from the respective British Ordnance Survey maps. However, it is also clear that

3.4 The Royal Naval Dockyard at Pembroke Dock, shown on Soviet plan of *Pembroke*, UK, 1:10,000, 1950.

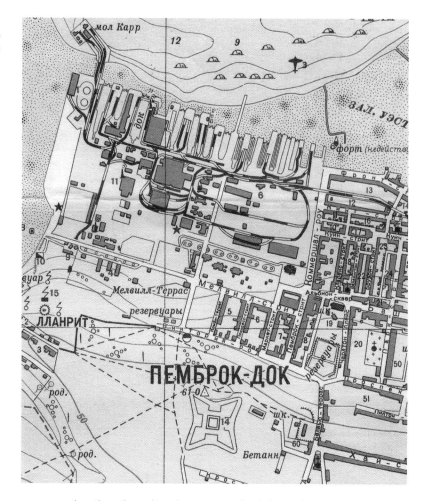

sources other than those listed were consulted. A good example is the 1950 plan of Pembroke on the Welsh coast, showing a Royal Navy dockyard, which later became an RAF flying boat base. The Soviet plan (fig. 3.4) was compiled in 1949 and cites 1:10,560 maps of 1934–36 as sources. In fact, the only relevant Ordnance Survey maps of that scale are dated 1869 (fig. 3.5), 1909 (fig. 3.6), and a 1948 survey, published in 1953 (fig. 3.7). The Soviet plan includes details redacted from these, such as the star-shaped "Defensible Barracks" and details of the buildings, interconnecting railways, and wharfs in the dockyard. The 1869 map shows the barracks and dockyard but no railways and no pier; on the later Ordnance Survey maps, the dockyard, pier, railways, and barracks are all omitted.

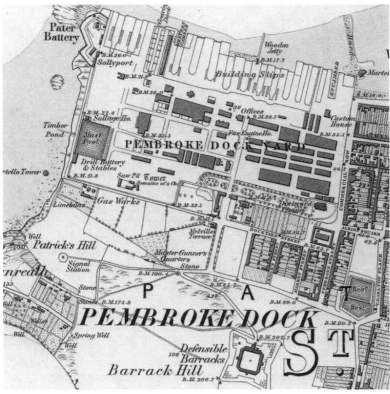

3.5 Ordnance Survey, Pembrokeshire sheet XXXIX, 1:10,560, 1869.

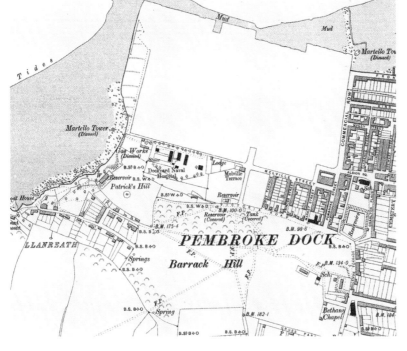

3.6 Ordnance Survey, Pembrokeshire sheet XXXIX.NE, 1:10,560, 1909.

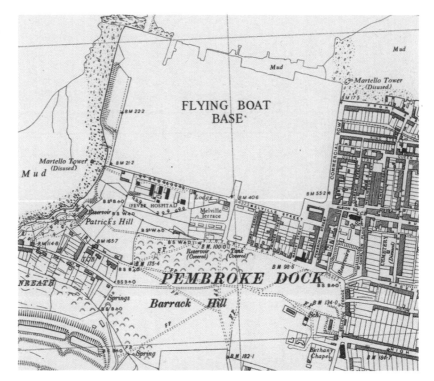

3.7 Ordnance Survey, Pembrokeshire sheet XXXIX.NE, 1:10,560, 1953.

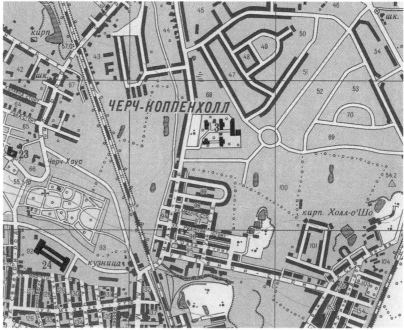

3.8 Part of Crewe, UK, 1:10,000, 1957, showing numbering of blocks

Many of the Soviet city plans of the 1950s and 1960s also have each block of land uniquely numbered. These numbers are not derived from local maps and seem to have been added by the cartographers for ease of identifying particular zones in the city. Figure 3.8 shows an example from Crewe, UK, where the numbers 47 to 54, for example, can be seen top right (see also fig. A1.6).

SPY IN THE SKY

Later city plans do not generally reveal the source mapping, although there are exceptions, such as that of Chicago: "Compiled from maps scale 1:24,000 of 1960, 63, 65 and using materials of 1972, printed 1982"; and an exceptionally late example, Vancouver: "Compiled from 1:25,000 scale maps of 1959–1984 and 1:50,000 scale maps of 1961–1981 using materials dated 1989, printed 2003."

By the late 1960s, following the launch of the Zenit satellite program in 1962, increasing use was made of satellite reconnaissance and imagery (see fig. 1.2). From then on, the large-scale city plans use these images, supplemented by a wealth of other detail derived from multiple sources of many different dates.

As will be seen, the primary cartographic source material in the United States is the US Geological Service (USGS) 7.5 × 15 minute mapping (at a scale of 1:24,000 or 1:25,000), and in Britain, the Ordnance Survey (OS) six-inch map (1:10,560 or 1:10,000). Both series provide national coverage and were current at the time that the Soviet maps were compiled. It is by comparing the Soviet maps with the latest national map at the time that the use of aerial reconnaissance can be proved.

One indicator of such use is where the map is demonstrably wrong as a result of misinterpretation of the imagery. The Teesside, northern England (fig. 3.9), and Los Angeles, California (fig. 3.10), city plans both show a road under construction. In the former case, at Thornaby-on-Tees, a gas distribution pipeline was being laid at the time, and the long linear excavation seen in the satellite imagery was evidently misinterpreted as road construction. In the latter case, in Granada Hills, San Fernando, the reason for the error is not apparent.

Another clear case of an error arising from misinterpretation of aerial imagery can be seen northeast of Boston, Massachusetts. The Soviet plan

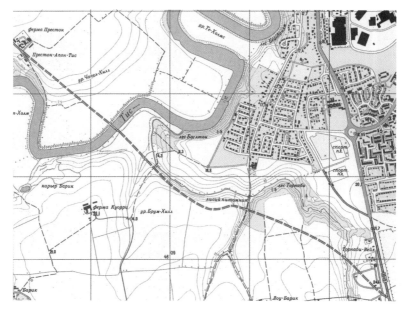

3.9 Erroneous road under construction on *Teesside*, UK, 1:10,000, 1975.

3.10 Erroneous road under construction at Granada Hills on *Los Angeles*, 1:25,000, 1976.

3.11 Junction of I-95 and US-3 at Burlington on *Boston*, 1:25,000, 1979.

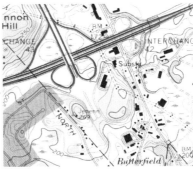

3.12 USGS, *Lexington*, Massachusetts, 1:25,000, 1971.

3.13 USGS, *Lexington*, Massachusetts, 1:25,000, 2015.

3.14 Non-existent road at Lexington on *Boston*, 1:25,000, 1979.

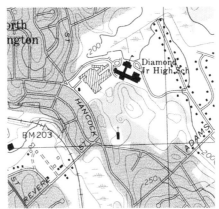

3.15 USGS, *Lexington*, Massachusetts, 1:25,000, 1971.

(fig. 3.11) has a symmetrical clover-leaf junction at the intersection of I-95 and US-3 near Burlington. The 1971 USGS map (fig. 3.12) shows a different road layout; but as the 2015 USGS sheet depicts (fig. 3.13), the earthworks and embankments were constructed for an intended symmetrical interchange and the extension of Route 3 southward, but the roads were never built.

The same Boston plan (fig. 3.14) shows a road running northeast, north of Diamond Junior High School, which the USGS sheet (fig. 3.15) reveals to be a watercourse. Although the evidence of the ditch on the US map was ignored by the cartographers, they did copy the street names—Revere, Hancock, and Adams (omitting those unnamed on the US map)—and the spot height (203 feet, equal to 61.9 meters).

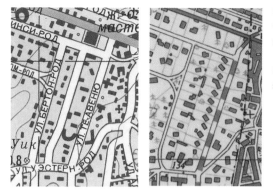

3.16 Non-existent road at Westbourne on *Bournemouth and Poole*, UK, 1:25,000, 1990.

3.17 Ordnance Survey, 1:25,000 sheet SZ09.

3.18 New housing at Opa-Locka on *Miami*, 1:25,000, 1984.

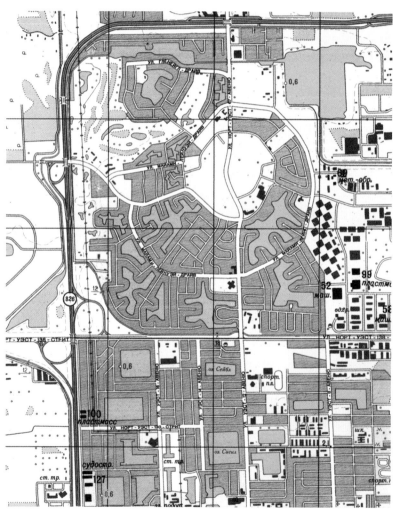

Another clear example where an aerial image has been misconstrued is at Westbourne on the 1:25,000 Bournemouth, UK, plan (fig. 3.16), on which an unnamed road is depicted running southwest to northeast. No such road exists, as can be seen from the Ordnance Survey map of the same scale (fig. 3.17).

The second indicator of the use of aerial images is where the Soviet plan shows details that are more recent than those depicted on the latest national map available at the date the Soviet city plan was compiled. An example is seen on the Miami, Florida, plan, which shows an extensive housing development at Opa-Locka (fig. 3.18) not shown on the then-latest USGS map (fig. 3.19).

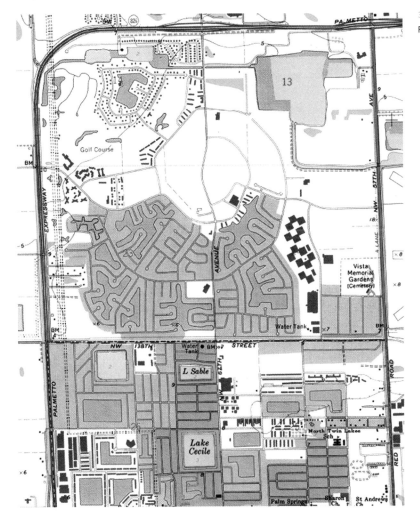

3.19 USGS, *Opa-Locka*, Florida, 1:24,000, 1973.

PLOTS AND PLANS

3.20 Part of Alexandria, Virginia on *Washington, DC*, 1:25,000, 1975.

3.21 USGS, *Alexandria*, Virginia, 1:24,000, 1965.

3.22 USGS, *Alexandria*, Virginia, 1:24,000, 1971.

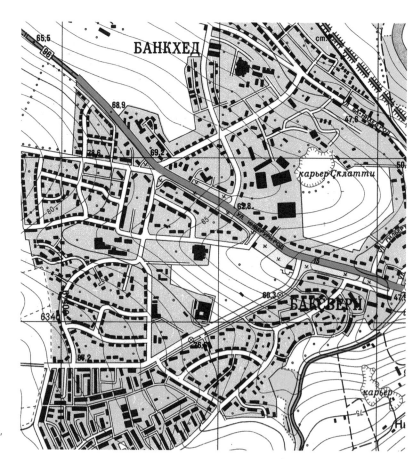

3.23 New housing at Bucksburn on *Aberdeen*, UK, 1:10,000, 1981.

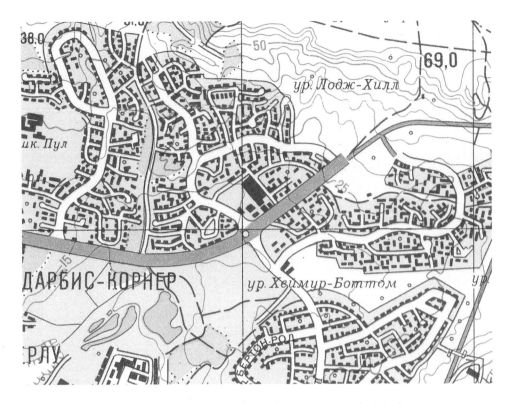

3.24 New housing at Canford Heath on *Bournemouth and Poole*, UK, 1:25,000, 1990.

A rather confusing situation is seen at Alexandria, Virginia, on the Washington, DC, plan (compiled 1973, printed 1975), figure 3.20, which shows information not yet portrayed on the latest USGS maps but also omits some developments that are on the US map. Figure 3.21 shows the 1965 US sheet and figure 3.22 the same revised in 1971 (changes in purple). The Soviet sheet shows a new housing area south of Huntington Avenue, but not the new L-shaped building opposite. It shows one new building south of the trailer park (labeled "*храм*" [church]) and the new road north of the railway, but not the other building by the trailer park, the one on the opposite side of Telegraph Road, or the new interchange at the north end of Telegraph Road.

Many of the city plans show areas of new housing developments with street names omitted, indicating that the layout has been taken from aerial images and that no local maps or directories were available to provide the names (if the names were known, they would have been shown). Examples are shown in figures 3.23 and 3.24. Another example can be seen in figure 3.9, in the area to the east of the river.

PLOTS AND PLANS

The use of aerial images is also implied where Soviet mapping (topographic and city plans) shows information that was deliberately omitted from national mapping for security reasons. In some cases, of course, an agent on the ground could be responsible for supplying some of the detail, but would be unlikely to adequately survey the entire site to establish the layout and size and shape of buildings.

One British example is the depiction of the munitions factory at Burghfield, the atomic weapons establishment where Trident-mounted nuclear warheads were manufactured, serviced, and maintained during the Cold War and later decommissioned. The *Guardian* newspaper of August 7, 2006, reported:

> Ordnance Survey has finally stopped falsifying Britain's maps, almost eighty years after the government first ordered cartographers to delete sensitive sites in the hope of thwarting German bombers. The popular 1:50,000 Landranger series will now show the nuclear warhead plant at Burghfield, near Reading, hitherto shown as a mysteriously empty field although well known to anti-nuclear demonstrators.

However, the site is shown in detail on the 1982 Soviet 1:50,000 sheet M-30-022-4, seen in figure 3.25. Strangely, Google Maps continues to omit the buildings or to identify the site, although they can be clearly seen on Google Earth.

3.25 Burghfield AWE (Atomic Weapons Establishment) on 1:50,000 sheet M-30-022-4, 1982.

3.26 Saughton Prison on *Edinburgh*, 1:10,000, 1983.   3.27 Ordnance Survey 1:25,000 sheet NT27.

Other sites redacted from Ordnance Survey maps (known as "security deletions") included prisons and sensitive dockyards. Typical examples include Saughton Prison, shown on the 1983 Edinburgh plan (figs. 3.26 and 3.27); the Royal Naval Dockyards at Chatham, where British submarines were built and maintained during the Cold War (figs. 3.28 and 3.29); and Plymouth (figs. 3.30 and 3.31). The Soviet plan of Chatham was compiled in 1982 and printed in 1984—ironically, the year that the yard closed. Interestingly, Chatham's Royal Naval Dockyard is color-coded green, denoting a military establishment, while that at Plymouth is black, denoting industrial premises.

The USAF base at Upper Heyford, Oxfordshire—from where bombers loaded with nuclear weapons were capable of flying missions over the USSR—was, however, shown on Ordnance Survey maps; the 1981 Soviet sheet M-30-010-1 (fig. 3.32) is evidently copied directly from this, as evidenced by the inclusion of the "steep hill" arrow symbol—standard OS notation, but meaningless on Soviet maps.

There are many American examples of the Soviet maps showing details omitted from (or out-of-date on) USGS mapping and therefore, presumably, derived from satellite imagery. Hanscom Field Air Force Base appears on the 1979 Boston plan (fig. 3.33), which shows details not shown on the latest USGS sheet (fig. 3.34), such as the perimeter security fence, the runway extension in the northeast corner, and the northwest–southeast runway. The Russian abbreviation "ВВС" indicates air force.

Similarly, the 1980 Soviet map of San Diego (fig. 3.35) shows specific individual buildings within the naval base that appear only as generalized

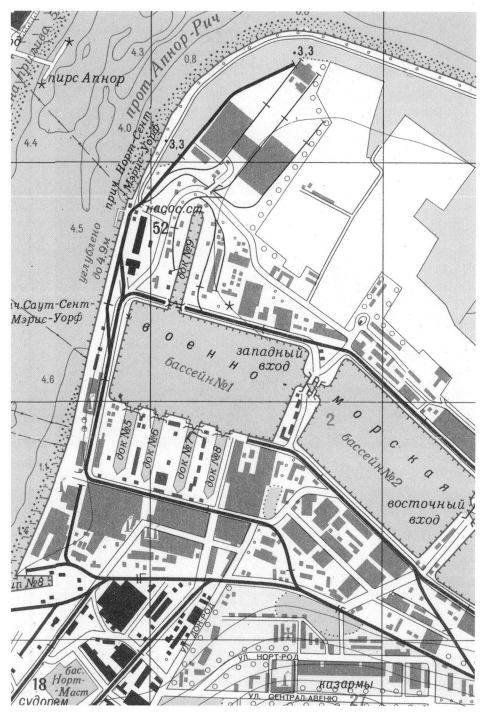

3.28 Royal Naval Dockyard Chatham on *Chatham, Gillingham, and Rochester*, UK, 1:10,000, 1984.

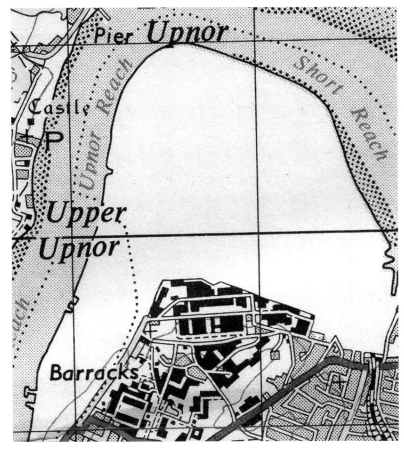

3.29 Ordnance Survey 1:63,360 sheet 172.

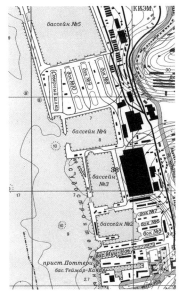
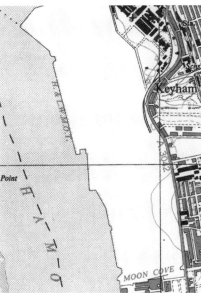

3.30 Royal Naval Dockyard Devonport on *Plymouth*, UK, 1:10,000, 1981.

3.31 Ordnance Survey 1:25,000 sheet SX45.

3.32 USAF Upper Heyford, UK, on 1:50,000 sheet M-30-010-1, 1981, showing "steep hill" arrow symbol (*top left*) copied from Ordnance Survey map.

3.33 Hanscom Air Force Base on *Boston*, 1:25,000, 1979.

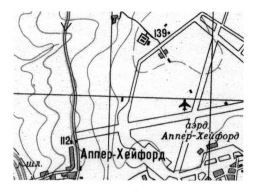

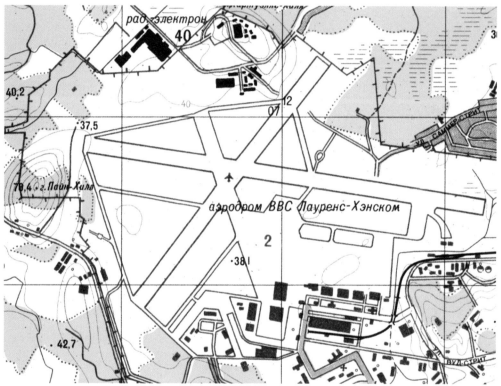

areas on the USGS map (fig. 3.36). More of the streets immediately east of the base are named on the Soviet map. The same two maps depict the nearby US Naval Training Center and the US Marine Corps Recruiting Depot, both adjacent to the International Airport (figs. 3.37 and 3.38). Again, areas shown as generalized blocks on the American map have the individual buildings

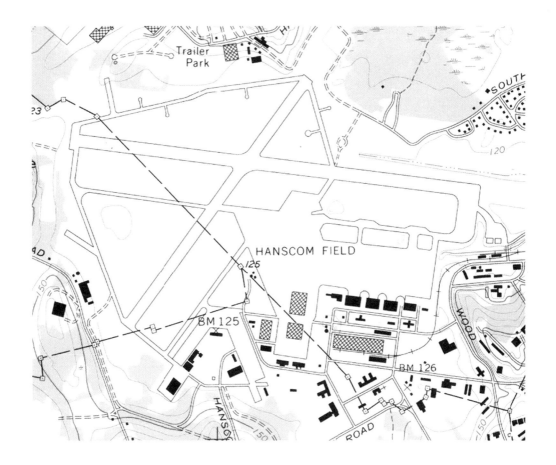

detailed. The green color-coding on figures 3.33, 3.35, and 3.37 indicates sites of military significance.

3.34 USGS, *Concord, Massachusetts*, 1:24,000, 1970.

The US Naval Reservation Bayonne Supply Center and the adjacent Greenville Yards, New Jersey, appear more fully developed on the 1982 Soviet plan of New York (fig. 3.39) than on the then-latest USGS map (fig. 3.40). The Soviet sheet shows a new jetty and buildings. The text at the new buildings reads: "Railway sorting and goods station Greenville." The abbreviation "BMC" means Navy; "ВМБ" is naval base.

At John F. Kennedy International Airport, New York, the Soviet plan of 1982 omits the by-then-disused runway running southwest–northeast shown on the latest USGS map and shows later development of the roads and interchanges to the northwestern corner of the site (figs. 3.41, 3.42).

PLOTS AND PLANS

65

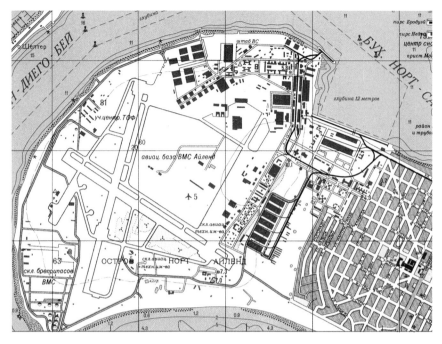

3.35 US Navy Base on *San Diego*, California, 1:25,000, 1980.

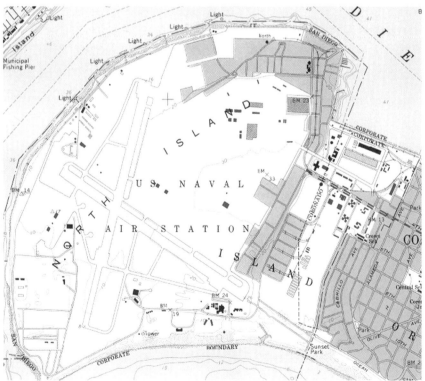

3.36 USGS, *Point Loma*, California, 1:24,000, 1975.

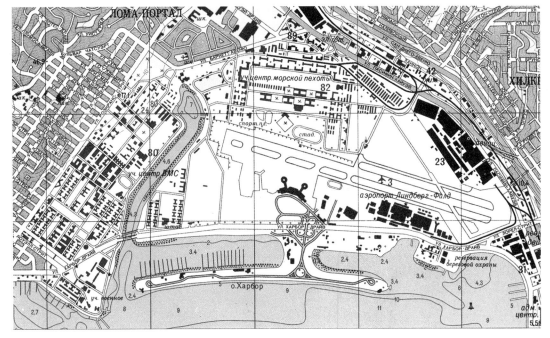

3.37 US Naval facilities on *San Diego*, California, 1:25,000, 1980.

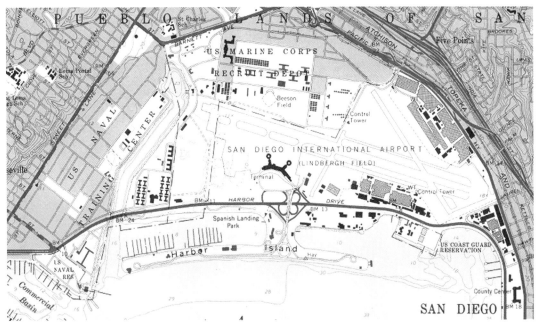

3.38 USGS, *Point Loma*, California, 1:24,000, 1975.

3.39 US Navy Bayonne and Greenville Yards, New Jersey on *New York*, 1:25,000, 1982.

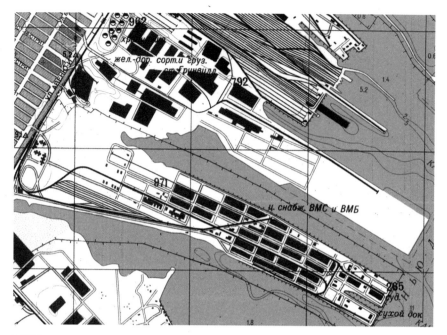

3.40 USGS, *Jersey City*, New Jersey, 1:24,000, 1967.

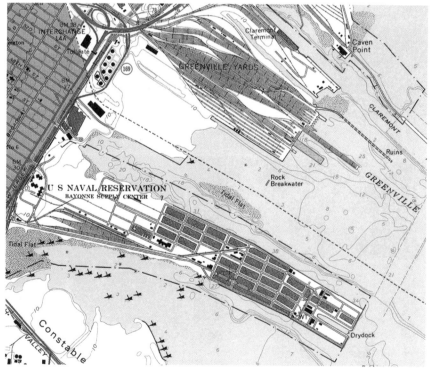

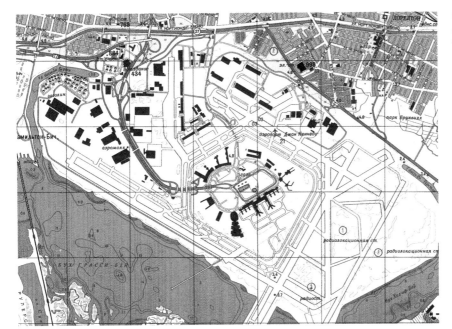

3.41 John F. Kennedy Airport on *New York*, 1:25,000, 1982.

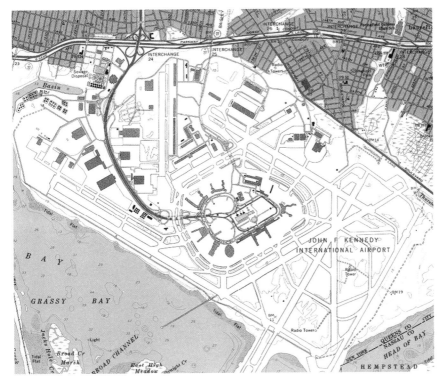

3.42 USGS, *Jamaica*, 1:24,000, 1966.

SECRET SECRETS

Omission of secret sites from national maps was no hindrance to the Soviet compilers in the case of installations that could be recognized from aerial surveillance. But "hidden" locations do seem to have escaped their attention. Examples include the eleven Regional Seats of Government in Britain, which were underground bunkers, established in the late 1950s to the mid-1960s, from which the country would be run in case of nuclear attack. The general population was unaware of their existence, and no evidence of their presence was visible on the ground or on published maps or documents.

Another example is the secret Cold War communications towers in Tenleytown, Washington, DC, although these were "hiding in plain sight" and would have been apparent to an observer on the ground, but evidently not recognizable from aerial imagery.

BOOTS ON THE GROUND

How large a part did personal observations play in the data gathering? A fascinating piece of evidence regarding secret agents collecting information on the ground comes from a story appearing in the English-language Russian news website *Pravda Online* in 2003 [16]. This describes the shock caused in Sweden by the publication in the Swedish newspaper *Aftonbladet* of illustrations of Soviet maps, dated 1987, of Stockholm and Karlskrona (the principal Swedish naval base). The detailed maps were of very high quality ("better than the creations of the best Swedish military mappers") and showed all defensive installations and depths of secret waterways. The story continues:

> The publication in the Swedish newspaper ruined two myths at once: that Swedish production is always of better quality against the Russian one, and Sweden managed to conceal very important information about its coastal defence from Russians. The published maps contained the information even about the berth length and the depth at secret naval bases, not to mention the location of secret mine fields. Christer Holm, a military intelligence chief [said] that the maps were most likely drawn on the base of secret agents' information.

Tore Foshberg, the retired chief of the counterintelligence department of the Swedish secret police, told reporters about the way the exposed maps were drawn up. Russian Central Intelligence Administration and KGB agents (there were up to 45 of such agents at the Soviet Union embassy in Stockholm) would go on a tour around the country. During their short journeys they would check loading capacities of bridges, they would also measure the distance between trees in a forest. Such information was necessary to plan the moves of the Soviet incursion army on the territory of Sweden. Soviet diplomats would arrange picnics in the places of strategic interest, they would be very friendly and sociable to the local population. "One of them, military attaché Pyotr Shiroky, went to a beach near Stockholm one fine day in the summer of 1982. On that beach he started a conversation with an excavator driver (as if incidentally), who was resting on the sand nearby. As it turned out, the driver dug trenches for the cables, which were connected to mine fields. However, there was a Swedish secret agent on the beach too. He heard the entire conversation," Tore Foshberg remembers.

What evidence is there of agents on the ground providing the information used in the Soviet maps of Britain and North America? While there *may* have been extensive use of agents to collect material and take observations or measurements, it is not easy to identify details that could *only* have come by this method. But there is one possible indicator—and the fact that it is a requirement of the global specification, but only appears sparsely on maps of the West, suggests that this is a reliable indicator of "boots on the ground." The indicator in question is the numeric annotation of specific attributes such as the load-carrying capacity of bridges; the clearance under a bridge; the speed of flow of a river; the typical girth, height, and spacing of trees in a forest; and various other measurements. This information is not shown on any national mapping and was not published elsewhere or discernible from aerial imagery; it requires personal on-the-spot inspection or estimation. Such annotation appears extensively on Soviet mapping of the USSR, where it is readily obtainable, but rarely on mapping of the United States and United Kingdom; thus where it does appear, it is reasonable to speculate that this particular place was chosen for a personal visit.

3.43 Medway River Bridge on *Chatham, Gillingham, and Rochester*, UK, 1:10,000, 1984.

3.44 Bridge at Eastern Shores on *Miami*, 1:25,000, 1984.

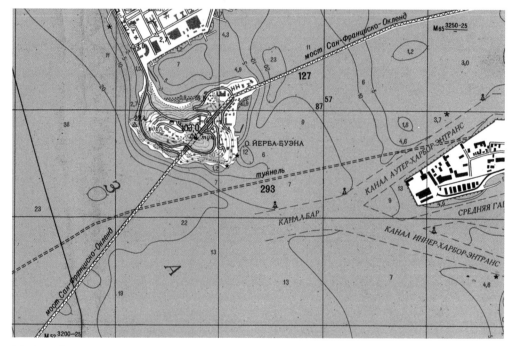

3.45 Extract from *San Francisco*, 1:25,000, 1980, showing the Oakland Bay Bridge annotated in two places as being of metal construction with a clearance of 52 meters and 65 meters, respectively (note also the BART railway tunnel, object number 293).

Two of the rare examples of bridge attributes appearing on British or American maps are shown in figures 3.43 and 3.44. While the former is adjacent to the Chatham Royal Naval Dockyard (shown in fig. 3.29), no obvious explanation comes to mind to account for a spy's close attention to a minor road at Eastern Shores on the northern outskirts of Miami. The Chatham bridge is annotated as constructed from reinforced concrete, with a 6-meter clearance above the river; it measures 300 meters long by 15 meters wide,

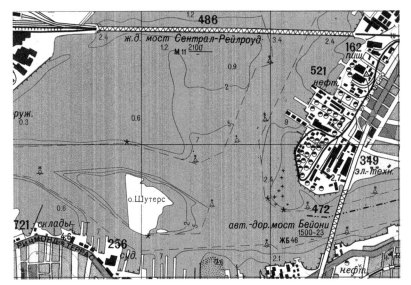

3.46 Extract from *New York*, 1:25,000, 1982, showing a metal railroad bridge (length 2,100 meters) labeled "Central Railroad bridge" and a concrete road bridge (1,500 meters long by 23 meters wide) named Bayonne Bridge.

with a load-bearing capacity of about 100 tonnes (metric tonnes; ~110 tons). The Miami bridge is also noted as being of reinforced concrete, 160 meters long and 15 meters wide, with a capacity of 7 tonnes.

Elsewhere, only dimensions that can be ascertained from aerial imagery are shown. For example, bridges in San Francisco (fig. 3.45) and Newark, New Jersey (fig. 3.46), are annotated with dimensions and construction material but not load-bearing carrying capacity. Annotation of road and carriageway widths appears extensively on major highways on UK maps (but generally not those of the United States)—again, readily established from aerial imagery. Figure 3.47 shows a typical example on the M11 motorway near London.

An unusual example of annotation is that of the width of minor roads in an area of new development west of Edinburgh (fig. 3.48); four instances (4, 5, 5, and 8 meters) appear in this small extract alone, with many more nearby; presumably the result of personal inspection by an agent on the ground. By contrast, figure 3.49 shows the extent of annotation that appears on plans of USSR territory, in this case Vilnius, Lithuania, and would have been expected to be shown on plans of the rest of the world, had the necessary freedom of access been possible. The symbols on this extract from the 1991 1:10,000 plan represent from west to east along the river:

PLOTS AND PLANS

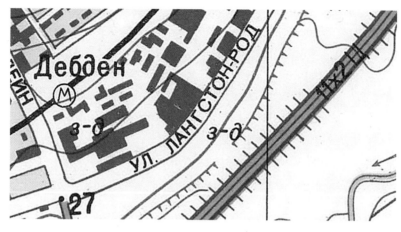

3.47 Extract from *London*, 1:25,000, 1985, showing the M11 motorway annotated as two 11-meter-wide concrete carriageways (note also the metro station, Дебден [Debden]).

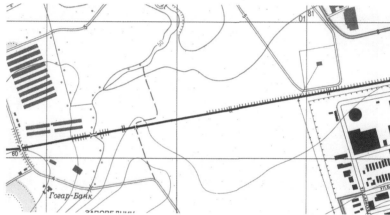

3.48 Annotation of widths of minor roads on *Edinburgh*, 1:10,000, 1983.

- a bridge constructed from reinforced concrete, 10 meters clearance above water, 105 meters long by 22 meters wide, with a carrying capacity of 30 tonnes;
- an embankment 5 meters high;
- an embankment 7 meters high;
- a river flowing to the west with a speed of 0.8 meters per second;
- a river name in uppercase letters indicating a navigable river (there are two river names as this is the confluence of the Neris and Vilnia);
- a concrete bridge 33 by 12 meters, with a carrying capacity of 50 tonnes;
- an elevation of a river surface 87.0 meters;
- a river depth of 1.71 meters with a sandy bed (п), river width of 88 meters;
- an embankment 6 meters high;
- a concrete bridge 175 by 14 meters, with a carrying capacity of 30 tonnes;
- an underpass 5 meters high and 6 meters wide.

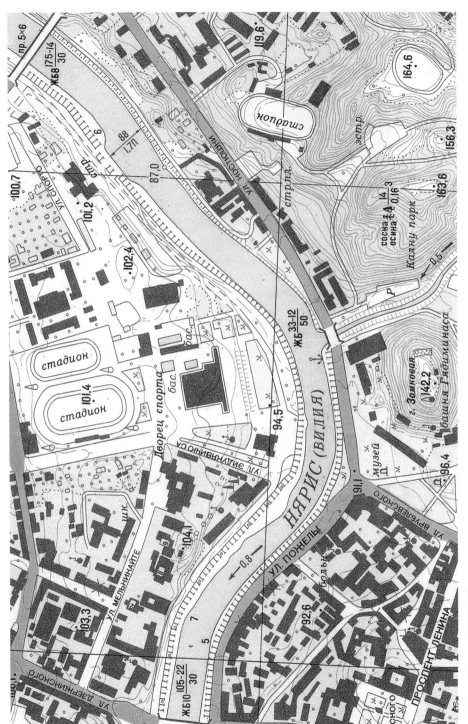

3.49 *Vilnius*, Lithuania, 1:10,000, 1991.

On the hill south of the river, the forest comprises mixed conifer and deciduous trees, named as "сосна" (pine) and "осина" (aspen), with an average height of 14 meters, girth of 0.16 meters, and spacing between trees of 3 meters, with the tributary river flowing at 0.5 meters per second. Unsurprisingly, it was impossible to collect this level of detail for cities in the Western world, which suggests, perhaps, that on-the-ground data gathering played a relatively minor role in the mapping of Western countries.

THE PAPER TRAIL

In order to prove that information has been taken from any specific source, it is necessary to identify "fingerprints" or unique features that could only have come from a particular map or document. Good examples of these are spot heights, names of districts, and mistakes or errors that could not have arisen other than by copying.

How High?

The best example of information that *must* have been derived from existing local maps is the depiction of spot heights (elevations). These are specific values shown at a specific location. Unlike a contour line, which may be deduced from stereo-photos, a spot height has been surveyed on the ground by a national mapping agency such as the OS or the USGS, as precisely as the instrumentation and the datum permit. From time to time, an area may be resurveyed, and revised spot height values may then appear on subsequent editions of the local map. Thus the presence of particular values will be a good indicator of the source map.

The Soviet plans of American cities almost always show the same spot heights as the contemporaneous USGS 7.5 × 15 minute (1:24,000) sheet, and where discrepancies occur they may well be the result of errors of copying or converting. In the extracts shown in figs 3.11 and 3.12, for example, the values appearing on the USGS map are 269, 172, and 200 feet. These convert to metric values of 82.0, 52.4 and 61.0 meters, respectively, which appear on the Soviet sheet. Similarly, in the Hanscom Field example, the USGS spot heights of 123 and 125 feet (fig. 3.34) convert to the respective values appearing on the Soviet plan (fig. 3.33)—37.5 and 38.1 meters.

In the plans of British cities, the general case is that the Soviet sheets from the 1970s to the 1990s show spot heights as depicted on Ordnance Survey County Series maps (six inches to the mile; 1:10,560) dated earlier than 1939. The later versions of OS maps generally give fewer spot heights, and many of these have new values. However, the correlation is not simple, as the following analysis shows. This study looks at a strip of approximately 4 by 2 kilometers (2.5 by 1.25 miles) across the northern part of the city of Bradford, West Yorkshire.

Table 3.1 shows all 14 spot heights within the strip shown on the 1990 Soviet plan compared with those on successive editions of OS six-inch maps. For ease of identification, the points have been labeled in red on figures 3.50 and 3.51. Figure 3.50 shows the western half, including points 1 to 8; figure 3.51 shows the eastern half, with points 9 to 14.

What the study reveals is that the cartographers combined data from three different contradictory maps (shown in boldface in table 3.1):

- the 1909 map for points 9, 10, 12, and 14
- the 1938 map for points 1, 4, 5, and 6
- the 1982 map for points 2, 3, 7, 8, and 11

Also point 13, which does not appear on any OS map, has been added by the cartographers by interpolation as the local highest point.

TABLE 3.1

Point	Soviet (m)	OS 1909 (ft.) (= m)	OS 1938 (ft.) (= m)	OS 1956 or 1968 (ft.) (= m)	OS 1982 (m)
1	134.5	441.6 (134.6)	441.22 (**134.5**)	—	—
2	175.0	—	—	575 (175.3)	**175**
3	168.0	553.7 (168.8)	553.3 (168.6)	—	**168**
4	189.9	623.3 (190.0)	622.93 (**189.9**)	622 (189.6)	—
5	234.8	768.5 (234.2)	770.2 (**234.8**)	—	243
6	253.6	831 (253.3)	832.05 (**253.6**)	830 (253.0)	253
7	248.0	—	—	—	**248**
8	226.0	—	—	713 (217.3)	**226**
9	195.4	641 (**195.4**)	—	642 (195.7)	196
10	204.0	669.2 (**204.0**)	668.78 (203.8)	665 (202.7)	—
11	197.0	—	—	—	**197**
12	104.1	341.7 (**104.1**)	—	339 (103.3)	—
13	208.0	—	—	—	—
14	139.6	458.3 (**139.7**)	—	—	—

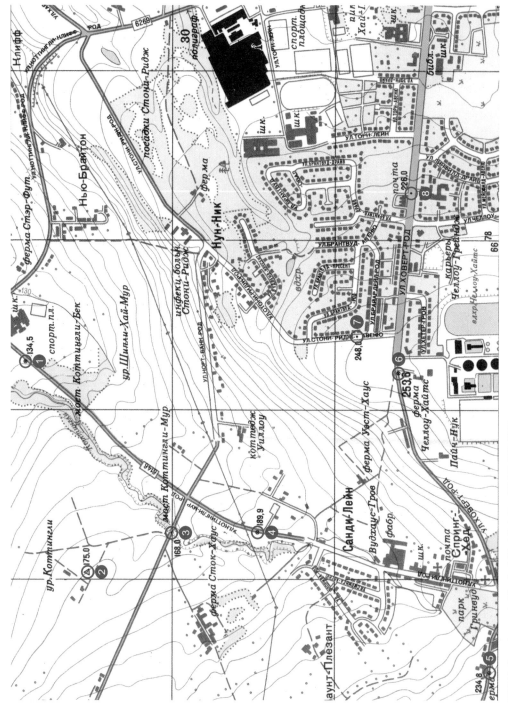

3.50 Northwestern part of *Bradford*, West Yorkshire, UK, 1:10,000, 1990.

3.51 Northern part of Bradford, West Yorkshire, UK, 1:10,000, 1990.

NAME THAT PLACE

Place-names (toponyms) may provide evidence of the use of particular source maps or other documentation. In some cases, the name attached to a district of a city may vary over time, or—particularly if a name is used informally locally—different versions may appear on different maps and street atlases. The 1990 Bradford plan again provides an intriguing example. An analysis of some of the district names appearing on the Soviet plan against six previous OS six-inch editions (1852, 1909, 1938, 1956, 1968, and 1982) and seven commercial street atlases (1961, 1972, 1973, 1980, 1982, 1988, and 1993) reveals the following remarkable results:

- two district names (Junction and Primrose Hill) appear only on the 1852 OS map;
- Dumb Mill Place appears only on the 1909 OS map;
- Fagley appears only on the 1982 commercial atlas;
- Lower Grange appears in two different locations on the Soviet plan, neither of which matches any local map;
- The Boggs does not appear on any of these local maps;
- Wyke Common appears on the Soviet plan located west of Wyke, but is shown east of Wyke on all local maps.

In addition, the name КОЛ-ПЛЕИС, "Coll Place" (fig. 3.52), appears twice on the Soviet sheet—once in the vicinity shown on the 1909, 1938, and 1956 maps by Halifax Road (ГАЛИФАКС-РОД) and once as shown on the 1967, 1988, and 1993 maps. The latter location is adjacent to Huddersfield Road, which is wrongly labeled "Manchester Road" (МАНЧЕСТЕР-РОД). Clearly, the cartographer became confused by the bewildering array of inconsistent material available to work from. Analyses of several other maps give similar results, if not quite so dramatic. By contrast, on the plan of Dublin, Ireland (compiled 1970, printed 1980), every district name can be traced directly to one map, the Dublin street plan produced in 1948 by the Ordnance Survey of Ireland.

One interesting feature about the naming of places on Soviet maps is that the local name is rendered phonetically in Cyrillic. Thus, an English name such as Leicester is transliterated as Лестер (Lester) and Gloucester as Глостер (Gloster). This may be helpful if the map user needs to ask directions from a native, but would be unhelpful if trying to navigate using road signs. What is

3.52 Coll Place on *Bradford*, West Yorkshire, UK, 1:10,000, 1990.

3.53A Keynsham on *Bristol*, UK, 1:10,000, 1972.

interesting is that the local pronunciation is almost always correctly given, even for rather obscure English names such as Wymondham in Norfolk, correctly rendered as Уиндем (Windem) on sheet N-31-123.

The convention is followed for the transliteration of other places around the world. For example, Marseilles (France) is spelled Марсель; The Hague (Netherlands), Гаага; Tehran (Iran), Тегеран; Suez (Egypt), Суэц; Badajoz (Spain), Бадахос; and Tangiers (Morocco), Танжер. In some cases, alternative pronunciations are offered in brackets; examples include Keynsham, appearing on the Bristol city plan as "КИНШАМ (КЕЙНСЕМ)," roughly "Kinsham (Keynsem)" (fig. 3.53A). An added complication in the Republic of Ireland is that many places have both a Gaelic Irish and an English-language name. Figures 3.53B and C show two small towns about a mile apart in County Carlow. In both cases, the Soviet cartographers have attempted to capture the correct pronunciation of both names. In the latter case, both the Irish and English versions are actually incorrect, but in general across Ireland the names are accurately represented. Help to achieve this could have come from the Russian Embassy in Dublin or the language department at Moscow University, where Gaelic Irish was taught. The Soviet cartographers would have been comfortable with the challenge of different languages and their various scripts, there being some 133 languages or dialects spoken within the territories of the USSR at that time.

The city of Kingston upon Hull, so-named on Ordnance Survey maps but commonly called just Hull, appears on the 1:100,000 sheet N-30-084 as "Гулль

3.53B 1:100,000 sheet N-29-119, *Carlow*, Ireland, 1979, showing the town name "Leighlinbridge" followed by its Irish alternative "Leithghlinn an Droichid," both rendered approximately correctly phonetically.

3.53C 1:100,000 sheet N-29-119, *Carlow*, Ireland, 1979, showing the town name "Muine Bheag" followed by its English alternative, "Bagenalstown," both incorrectly rendered. МУИМ-БАГ, pronounced "Muim Bag," should be "Mwina Vug," and БАДЖЕНАЛСТАУН, or "Badgenalstown," should have a hard "g" as in "bagel."

(Халл)"—both pronounced approximately "Hull." Nearby Goole appears on N-30-083 as Гул, although this time the initial G is hard in English. The Leeds plan reveals a related difficulty; the districts of Gipton and Gildersome are both pronounced with a hard G, yet are named on the plan with Cyrillic initial letters ДЖ, pronounced as a soft G (DJ). Despite the occasional error, all of

3.54 Underlined name, Nitshill, on *Glasgow and Paisley*, Scotland, 1:25,000, 1981.

this indicates an impressive amount of research and geographic knowledge—whether or not it implies local input is impossible to judge.

One characteristic of place-names is that multiple-word names are always hyphenated on Soviet maps, whereas not shown so on native British or American maps. Examples seen above include the following:

- Пемброк-Док, Pembroke Dock (fig. 3.4)
- Вудс-Корнер, Woods Corner (fig. 3.11)
- Аппер-Хейфорд, Upper Heyford (fig. 3.32)
- Истерн-Шорс, Eastern Shores (fig. 3.44)
- Кол-Плеис, Coll Place (fig. 3.52)

An interesting attribute relating to place-names is the presence of underlined names. A few examples appear on the large-scale city plans of British cities, such as Nitshill (НИТСХИЛЛ), a suburb of Glasgow (fig. 3.54), as well as the suburbs of Manchester, Sheffield, and elsewhere; none have been seen on those of US cities. The significance of the underlining is to indicate that the name also applies to a nearby railway station.

LOST IN TRANSLATION

The errors or indicators that tie a map to a particular source also reveal the cartographers' problems in producing maps of a country that they have no experience with and no insight into cultural norms. The Soviet Doncaster (UK) plan, for example (fig. 3.55), gives the name РОМАН-ПОТТЕРИ-КИЛИС (Roman-Pottery-Kilns), to the Cantley estate, south of the town. This is indeed the site of extensive Romano-British potteries and is labeled as an archaeo-

3.55 "Roman Pottery Kilns" on *Doncaster*, UK, 1:10,000, 1976.

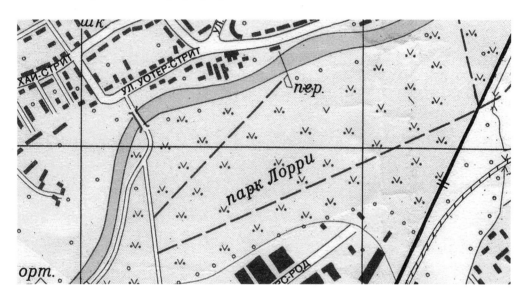

3.56 "Lorry Park" on *Cambridge*, UK, 1:10,000, 1989.

logical site on the Ordnance Survey six-inch maps. This has evidently been misinterpreted by the cartographers of the Soviet plan, who also confused the English letter N with the Cyrillic letter И—"Kilns" should be "КИЛНС."

There are other revealing examples of confusion or misinterpretation affecting items copied from maps or other sources. Both the 1977 and the 1989 editions of the Cambridge plan have an open green field by the river, just north of the city center, labeled "*парк Лорри*" (Lorry Park) (fig. 3.56). This space is named on OS maps and commercial street atlases as Stourbridge Common. It is hard to imagine how the compilers got the name wrong (possibly they

3.57 "Court" on *Cambridge*, UK, 1:10,000, 1989.

obtained a parking guide for truck drivers, although that sounds unlikely), and how the wrong name was carried forward to the second edition, despite all the other evidence they must have had in their possession, remains a mystery. Figure 3.57 shows a second mistake, also appearing on both editions of the Cambridge plan, the confusion over the word "court." Harvey Court, which is a building forming part of Gonville and Caius College, is wrongly identified as "*суд*" (courthouse, as in law courts).

On the Huddersfield plan (fig. 3.58), a small building in the tiny moorland village of Sowood Green, high in the Pennine Hills, is identified as object number 23 and listed in the index as "Институт технический" (fig. 3.59), or Institute of Technology, which is something that might be expected in a large city rather than a remote village. What it actually is (or was), as shown on the contemporaneous OS map (fig. 3.60), is the Mechanics' Institute, one of many educational establishments founded in the mid-nineteenth century for the betterment of the working man—to provide libraries and enlightenment

PLOTS AND PLANS

3.58 Extract from *Huddersfield*, UK, 1:10,000, 1984.

23   Институт технический

3.59 Index entry, *Huddersfield*, UK, 1:10,000, 1984.

3.60 Ordnance Survey, Yorkshire, UK, sheet CCXLV.SE, 1:10,560, 1949.

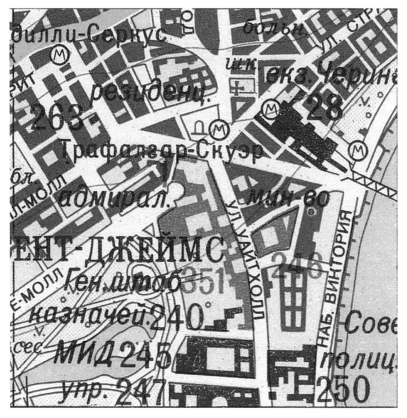

3.61 Her Majesty's Theatre (object 263) on *London*, 1:25,000, 1985. This extract is included within the larger area shown in figure A1.16.

for workers and to give them an alternative to spending evenings in the local pub. It is not surprising that a Soviet cartographer in the 1980s failed to appreciate the significance of such a culturally specific item of 150 years earlier.

Another cultural misunderstanding results in an amusing mistake shown in figure 3.61 from the plan of London, on which the West End theater, known as Her Majesty's Theatre (object number 263, upper left), is identified in the index as "Резиденция королевы и премьер-министра" (Residence of the Queen and Prime Minister).

A careless error can be seen on the Lancaster city plan (fig. 3.62; lower left), where a wayside inn, named on OS maps as the Golden Ball, is labeled "*гост. Болден-Голд-Инн*," or "Bolden Gold Inn." The depiction on the 1974 Liverpool plan of a "hydro-aerodrome" in the river Mersey at Bromborough

PLOTS AND PLANS

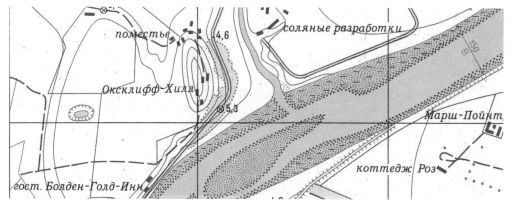

3.62 "Bolden Gold Inn" on *Lancaster and Morecambe*, UK, 1:10,000, 1983 (note also the nearby river Lune annotated as 150 meters wide and B indicating a viscous riverbed).

3.63 "Hydro-aerodrome" at Bromborough on *Liverpool*, UK, 1:10,000, 1974.

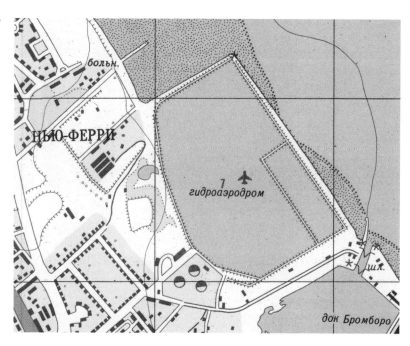

(fig. 3.63) provides probable evidence that the Russian cartographers had possession of captured German World War II invasion maps of Britain. An experimental flying boat service to Belfast operated from here briefly in 1928, which seems to be have been the only commercial use of the station. The facility appeared on quarter-inch (1:253,440) aviation maps of 1934 and 1939 and on commercial Bartholomew maps, but never on large-scale Ordnance

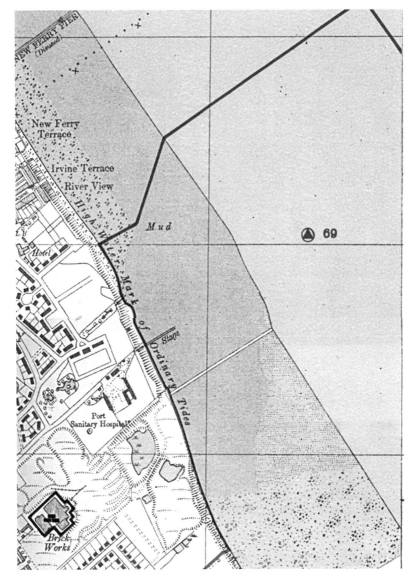

3.64 Generalstab des Heeres, Abteilung für Kriegskarten u. Vermessungswesen (German General Staff of the Army, Department of War Maps and surveying), BB12ai, Liverpool, UK, 1:10,000, 1942.

Survey maps. It does, however, appear on the 1942 German 1:10,000 plan (fig. 3.64), a photo-enlargement of the local OS six-inch sheet, with the addition of objects of military interest superimposed in red. Object number 69, shown in the river Mersey, is listed as "Flughaven" (airport). This is likely to be the source relied on by the Russian compilers.

Diligence in collecting information not shown on British maps is revealed

3.65 NATO Headquarters (object 352) on *London*, 1:25,000, 1985.

3.66 Headquarters of US Navy in Europe (object 350) on *London*, 1:25,000, 1985.

in three locations on the London plan. Figure 3.65 shows object number 352, which is identified as "Штаб Объединенных Вооруженных Сил НАТО" (Headquarters NATO Allied Powers), while nearby (not shown in the extract) is object 346, "Центр Оперативный ПВО НАТО Станмор" (NATO Air Defence Operations Centre, Stanmore). Figure 3.66 shows object number 350, listed as "Штаб ВМС США в Европе" (HQ of US Navy in Europe).

NEW MAPS FOR OLD

Interestingly, the materials available to the cartographers compiling the large-scale plans seem not to include existing editions of Soviet mapping of the same locality. Several British towns were mapped more than once (possibly many were, but only a few cases are known for certain). The later map is evidently a new, different map rather than an update of the original sheet. Figures 3.67 and 3.68 show extracts from the 1975 and 1989 editions, respectively, of the plan of Halifax, West Yorkshire. The position of the intersection of the grid lines (which have identical geographic values) and the contour lines on the two plans relative to Russell Park (парк Рассел) show that the later plan was drawn from scratch. Similarly, figures 3.69 and 3.70—the 1973 and 1986 Luton plans, respectively—show how differently the town center was depicted in the two editions.

3.67 Intersection of grid lines at Russell Park on *Halifax and Sowerby Bridge*, UK, 1:10,000, 1975.

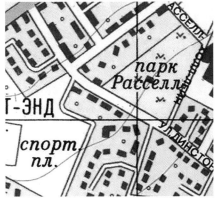

3.68 Revised location of intersecting grid lines shown on 1989 edition of *Halifax and Sowerby Bridge*, UK, 1:10,000.

3.69 *Luton*, UK, 1:10,000, 1973.

Other known instances of remapping include Bournemouth, Cambridge, Cardiff, Middlesbrough, Reading, and Wolverhampton. In all cases, the original mapping is at 1:10,000; for Bournemouth, Middlesbrough, and Wolverhampton, the remapping is part of 1:25,000 mapping of a larger area. No examples of remapping have been seen of cities in North America.

In parts of the UK where the distances between towns is relatively small, overlaps sometimes occur where a part of a conurbation is included in sev-

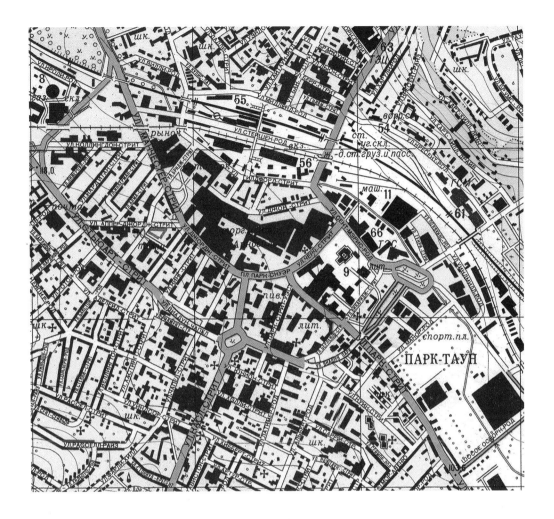

3.70 *Luton*, UK, 1:10,000, 1986.

eral surrounding city plans. An example is at Birkenshaw in West Yorkshire, which appears on the 1972 Leeds, 1983 Dewsbury, and 1990 Bradford plans (all 1:10,000), shown in figures 3.71, 3.72, and 3.73, respectively (road numbers in red have been added to figure 3.72 to aid the explanation). The three plans of this small area of about one square mile (2.5 km²) show considerable differences, as detailed below.

Street names: The Leeds plan has no street names while Dewsbury has six street names, two of which are incorrect: A651 is wrongly named Oxford Road (should be Bradford Road), and A652 is wrongly named Bradford Road (should be Dewsbury Road). The Bradford sheet repeats the Oxford Road

PLOTS AND PLANS

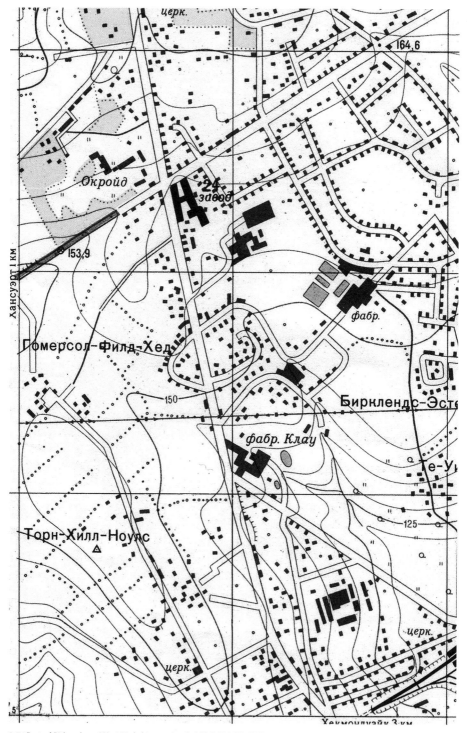

3.71 Part of Birkenshaw, West Yorkshire, on *Leeds*, UK, 1:10,000, 1972.

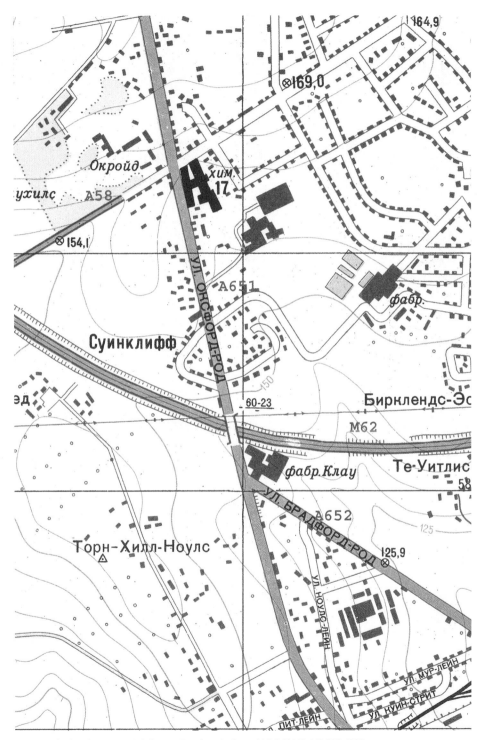

3.72 Part of Birkenshaw, West Yorkshire, on *Dewsbury, Batley, and Mirfield*, UK, 1:10,000, 1983.

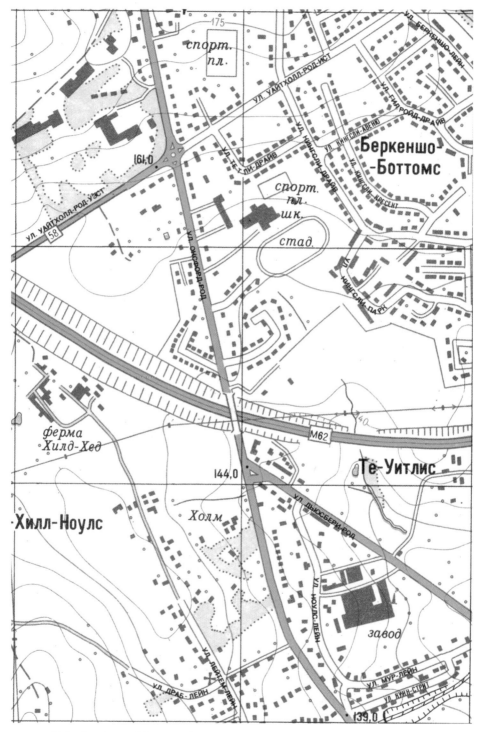

3.73 Part of Birkenshaw, West Yorkshire, on *Bradford*, UK, 1:10,000, 1990.

error, but names Dewsbury Road correctly, and twelve more names appear, one of which (Kingsley Park) is wrong.

Road color: The Leeds plan has A58 shaded up to the point where the built-up area starts, with other roads uncolored. Dewsbury also has the same stretch of A58 shaded, plus the full length of the other major roads. Bradford is similar, except A58 is shaded to junction with A651, but with the road casing ending at the start of the built-up area.

Spot heights: The Leeds plan has only 153.9 and 164.6 meters on A58. Dewsbury shows these points as 154.1 and 164.9 meters, respectively, and additionally has 169.0 meters on A58 and 125.9 meters on A652. The Bradford plan omits all of these, but has three different points—161.0, 144.0, and 139.0 meters—from north to south. The Leeds and Dewsbury plans show a triangulation point (or geodetic station) triangle symbol at Thorn Hill Knowles in the southwest.

M62 motorway: Not built at the time of the Leeds plan, the cutting is shown much wider on the Bradford plan than on that of Dewsbury.

Grid lines: There are slight differences on all three plans. For example, the central horizontal grid line runs through the southeast corner of the square pond on the Leeds sheet and on the northwest corner on the Dewsbury plan; on the Bradford plan, the area is redeveloped and the ponds are not shown. The grid lines' intersection adjacent to the Y-junction of A651 and A652 is in a slightly different position on each plan.

Woodlands (green-shaded areas): The shape of the woodlands northwest on the A58/A651 junction is different on all three plans. The Leeds and Dewsbury plans have no other green shading; the Bradford plan also shows two areas to the west of A651 and one south of M62.

OTHER AREAS OF OVERLAP ON UK PLANS INCLUDE THOSE FOR Sunderland and Newcastle, Bradford and Halifax, Huddersfield and Dewsbury, Manchester and Warrington, Wigan and St. Helens, Portsmouth and Havant, London and Thurrock. In addition to these overlaps, Liverpool abuts St. Helens, and Halifax abuts Huddersfield.

Similar contradictions occur with the 1:50,000 topographic maps, whose compilers seem not to have had access to existing larger-scale city plans. This can be seen across the southern part of England, where a new Soviet

3.74 Part of Paignton, Devon, on 1:50,000 sheet M-30-054-3, 1981.

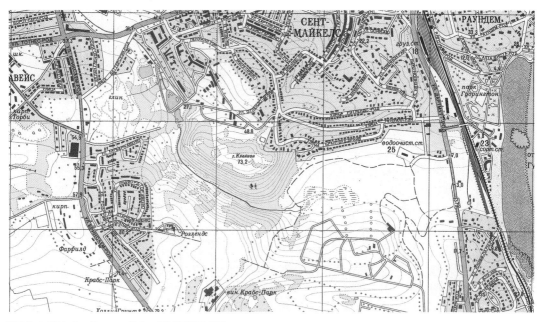

3.75 Part of Paignton, Devon, on *Torbay*, UK, 1:10,000, 1976.

1:50,000 series was produced in 1981, covering many towns for which city plans had been produced in the preceding ten or so years. Figures 3.74 and 3.75 show a typical example, in part of Paignton, Devon, where the later small-scale map omits a housing development (lower left) that had already appeared on the earlier (1976) large-scale Torbay plan.

TOWNS AND TRANSPORT

The cartographers' attempts to portray the required level of information in urban areas and transport networks seemed to have met with varying success

Плотно застроенные кварталы (участки кварталов):
а)-с преобладанием многоэтажных массивных зданий;
б)-с преобладанием малоэтажных мелких строений.

3.76 Marginalia on *Boston*, 1:25,000, 1979.

3.77 High-rise and low-rise blocks differentiated on *San Francisco*, 1:25,000, 1980.

PLOTS AND PLANS

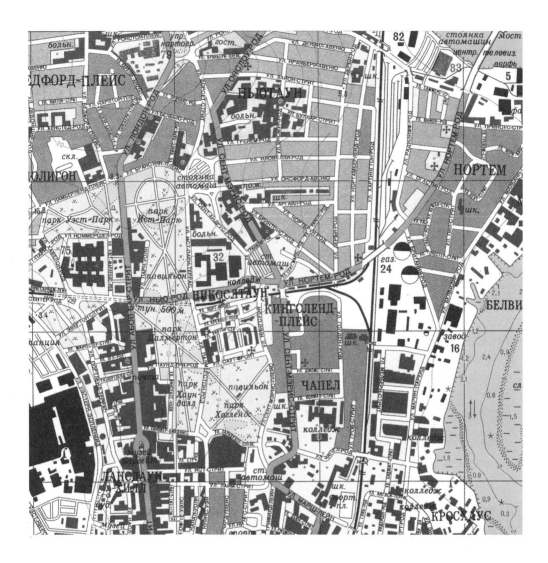

3.78 High-rise and low-rise blocks differentiated on *Southampton*, UK, 1:10,000, 1986.

and reveal again the need for data beyond that depicted on native national mapping or aerial imagery. Examples that follow show some cases of the success and failure in obtaining sought-after information.

The Soviet specification (fig. 3.76) calls for a distinction to be made between areas of generally low-rise and generally high-rise buildings. The key in figure 3.76 translates as "Built-up neighborhoods (areas of blocks): a) with a predominance of large multi-story buildings; b) with a predominance of small low-rise buildings." The Soviet maps of the United States generally do show this distinction, such as in the San Francisco plan (fig. 3.77); but in Britain the only example of this seems to be on the Southampton plan (fig. 3.78). Note the distinction between generalized areas of high-rise (medium brown), areas of low-rise (light tan), and footprints of individual buildings or blocks (dark brown).

RAILWAYS, METROS, AND STREETCARS

The railway network played a vital role in the development of the USSR, and the importance of railways and their significance in the functioning of society was evidently one of the cultural norms in the minds of the cartographers. The kind of information required to be gathered that was beyond what was shown on Western maps included the following:

- the distinction between electrified and non-electrified lines
- the number of tracks
- the location of station buildings relative to the tracks
- the dimensions of tunnels
- position of signal gantries (e.g., semaphore arms)
- the distinction between lines and stations forming part of a local metro system and those that were part of the national network

Considerable effort must have gone into collecting this information from a range of guides, directories, local transit maps, and photographs, as it can be seen that it was comprehensively (and generally correctly) portrayed on the large-scale Soviet plans of Great Britain and the United States.

Figure 3.79 shows a part of Wembley, London, with the non-electrified line running east to west and a station (Wembley Hill) on the right having the station buildings depicted to the south. This line crosses an electrified line, which has two stations (top and bottom of extract) served by both metro and national rail services. Figure 3.80 shows three signal gantries alongside the line near York, while figure 3.81 shows Totley tunnel, in Sheffield. The tunnel is object number 70 and is annotated as 5,200 meters long, with depth, width, and height not known (in fact, the annotation is inverted; the 5,200 should be above the horizontal, with the width and height below, and the depth alongside).

One piece of information that appeared on US national mapping but was

3.79 Annotation of railways on *London*, 1:25,000, 1985.

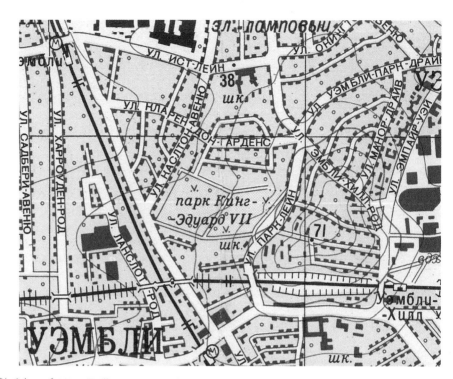

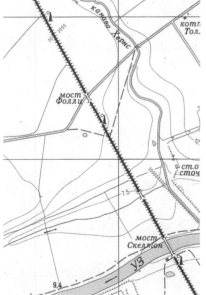

3.81 Annotation of Totley tunnel on *Sheffield*, UK, 1:25,000, 1977.

3.80 Railway signals on *York*, UK, 1:10,000, 1980.

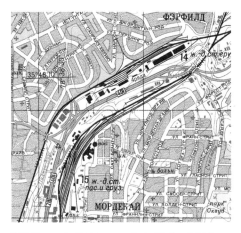 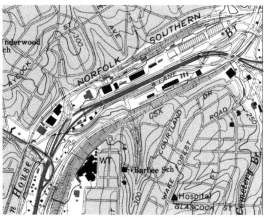

3.82 Names of railroad companies omitted on *Raleigh*, North Carolina, 1:25,000, 1980.

3.83 Railroad company names appear on USGS, *Raleigh West*, North Carolina, 1:24,000, 1988.

ignored by the Russian cartographers is the name of railroad companies. Figures 3.82 and 3.83 show the station at Raleigh, North Carolina, as depicted by the Soviet and USGS maps. The Soviet plan omits the "Norfolk Southern" and "CSX" labels alongside the railway tracks, but does have the "4 lane" label (4 ряда) alongside the highway. Note also the different coloring of the roads at the intersection where the railway crosses the road.

It is also notable that lines and stations that had been closed by then continue to be shown, even where anachronistic with the surrounding detail, showing that up-to-date information is to hand. This may indicate an assumption that a disused railway can be brought back into service or may simply be the result of assembling disparate pieces of data of different ages and provenance. An example is seen in Glasgow, where the 1981 plan (fig. 3.84) shows an electrified railway line running southwest from Neilston station. This section of line was never electrified and was closed in 1962 and the track lifted in 1964, thereafter being shown as dismantled on OS mapping, such as the 1965 sheet (fig. 3.85). The section of line to the northeast of Neilston station was electrified in 1962 and remained in use.

Similar examples occur on several maps. The Bristol sheet, for instance, shows several lines abandoned years earlier; the 1971 Leeds sheet not only shows lines closed in the 1960s, but also depicts and names the Wellington terminus station, adjacent to City station, although it had closed in 1938;

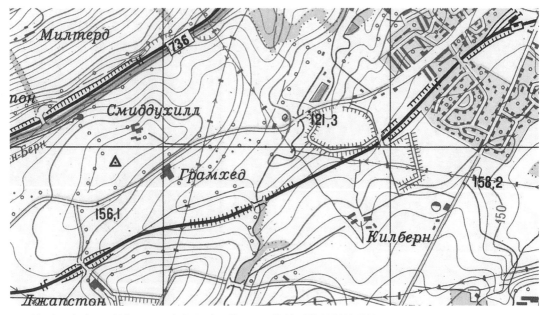

3.84 Abandoned railway at Neilston, wrongly depicted on *Glasgow and Paisley*, UK, 1:25,000, 1981.

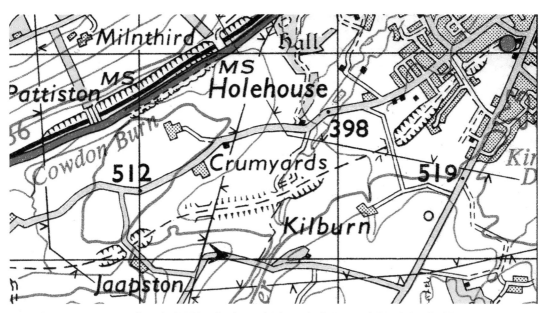

3.85 Ordnance Survey, 1:63,360, sheet 60, 1965 (Note that the spot heights on the Soviet map of 156.1, 121.3, and 158.2 meters are, as one would expect, the metric equivalents of those on the Ordnance Survey map: 512, 398, and 519 feet, respectively.)

3.86 Inset metro diagram on *Glasgow and Paisley*, UK, 1981.

while the 1983 Liverpool plan shows the Overhead Railway, which had been dismantled in 1957.

The treatment of metro systems (urban underground railways, otherwise known as subways, or "the tube" in England) is a little inconsistent—or possibly the definition of what constituted one was difficult to interpret. In Britain, only London and Glasgow qualified as having metro systems, and these are correctly portrayed with the M symbol at stations. US cities seem to have presented more of a problem; the Boston and New York plans each show metro systems; however, in some other cities the underground sections are ignored and the aboveground sections are shown as general railways.

The Glasgow (fig. 3.86) and Boston (fig. 3.87) plans have an inset diagram showing the layout of the system. Other city plans with similar diagrams include Liverpool (see fig. 2.21B), Rotterdam, and Stockholm. Figures 3.88 and 3.89 show underground and aboveground sections and the M symbol for stations

PLOTS AND PLANS

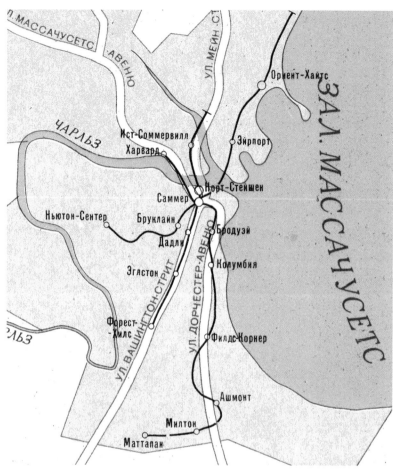

3.87 Inset metro diagram on *Boston*, 1979.

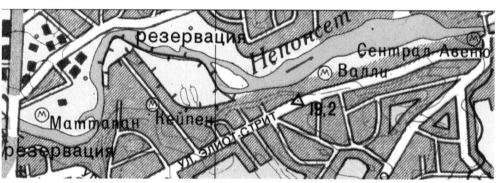

3.88 Underground metro stations on *Boston*, 1:25,000, 1979.

3.89 Aboveground metro stations on *Boston*, 1:25,000, 1979.

on the Boston metro at Mattapan and Milton, respectively. Figure 3.90, New York City, shows several M stations in close proximity in Lower Manhattan.

In Chicago (fig. 3.91), the city center elevated railway is not depicted as a metro system, although it serves a similar purpose and might have been expected to be identified as such. In San Francisco (fig. 3.92), a short aboveground section of metro at Mission is shown, with the rest of the network omitted. In Washington, DC (fig. 3.93), the absence of metro stations is explained by the compilation date (1973) preceding the opening of the system in 1976.

The London plan has a separate accompanying index booklet, and the metro diagram is a foldout supplement to this. Notably, the diagrams are geographically correct, unlike the familiar "tube map" in which the network is stylized as intersecting straight lines, distorting distances and relative locations. The London diagram has a number of errors, including transposed stations, misspelled names, and non-existent sections of line and

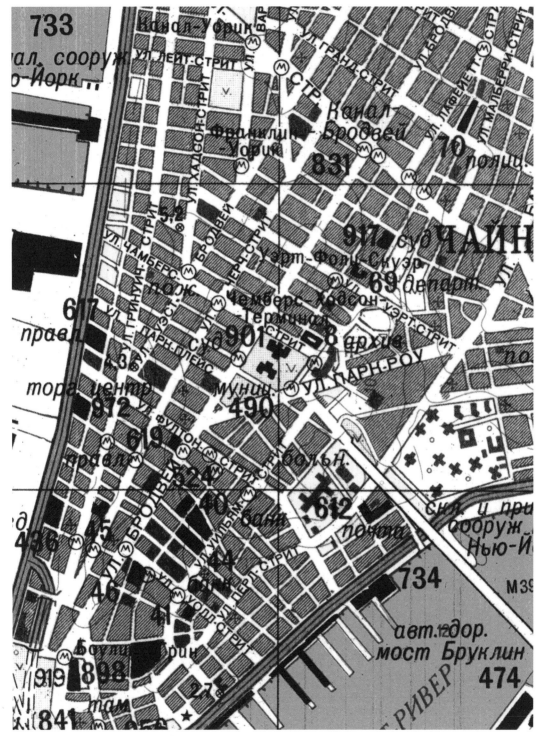

3.90 Metro stations in Manhattan on *New York*, 1:25,000, 1982.

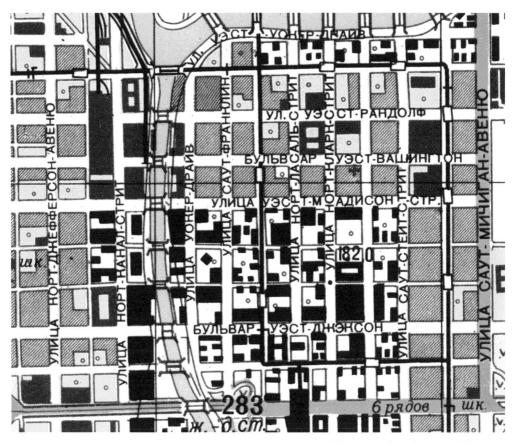

3.91 City center elevated railway on *Chicago*, 1:25,000, 1982.

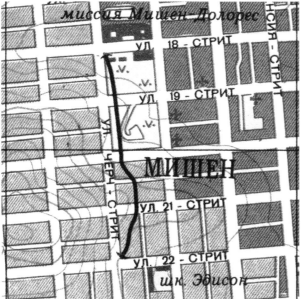

3.92 Aboveground section of metro on *San Francisco*, 1:25,000, 1980.

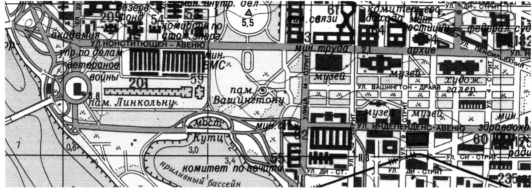

3.93 *Washington, DC*, 1:25,000, 1975.

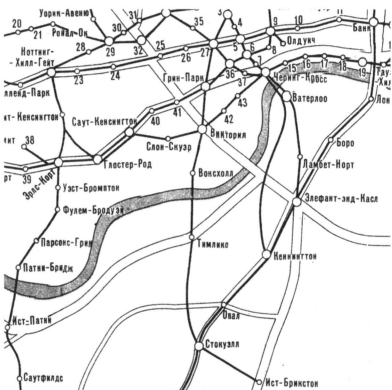

3.94 Part of diagram of metro lines on foldout supplement, *London*, 1985.

interchanges—possibly arising from trying to draw a geographic diagram of a complex network from the simplified official tube map. Part of the London diagram is shown in figure 3.94. Some of the errors that appear here include Brixton misnamed "East Brixton" (Ист-Брикстон), Pimlico misnamed "Timlico" (Тимлико), with its location transposed with Vauxhall (Воксхолл).

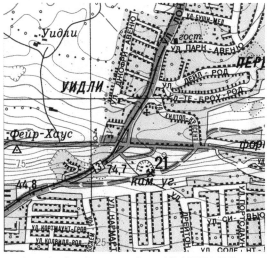

3.95 Long-abandoned tramway on *Havant*, UK, 1:25,000, 1983.

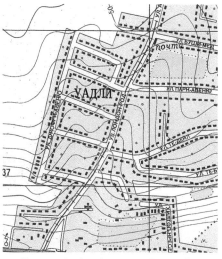

3.96 The same abandoned tramway on *Portsmouth, Fareham, and Gosport*, UK, 1:10,000, 1988.

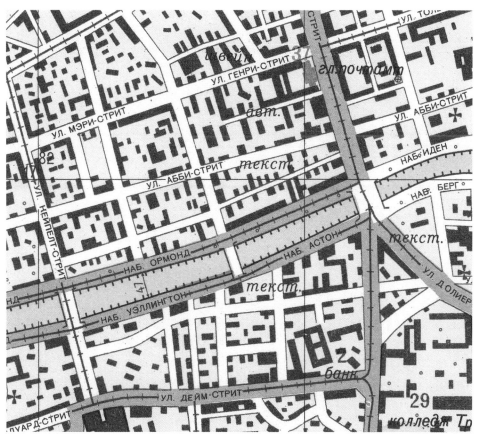

3.97 Long-abandoned tramways on *Dublin*, 1:10,000, 1980.

3.98 Cable cars omitted on *San Francisco*, 1:25,000, 1980. 3.99 USGS, *San Francisco North*, 1:24,000, 1956.

(Vauxhall is south of the river Thames.) Incidentally, the Russian word for "railway station," вокзал (pronounced "vok`zal"), is derived from this former terminus station and pleasure gardens.

Tramways (streetcars or urban light rail running at street level) are shown inconsistently. In Britain and Ireland, these had been completely abandoned and replaced by buses by the time of the Soviet mapping. However, both the 1983 Havant and 1988 Portsmouth plans (on an overlapping portion) show a tramway running along the main road connecting the towns (figs. 3.95 and 3.96, respectively). This service had closed fifty years earlier, in 1934. Similarly, the 1980 Dublin plan (fig. 3.97) shows an extensive tramway network throughout the city, although the system was gradually abandoned during the 1940s and finally closed down in 1949. Conversely, the famous San Francisco cable cars do not appear on the Soviet plan (fig. 3.98), where they might be expected. They are indicated on the then-latest USGS mapping (fig. 3.99) by the label alongside the relevant streets.

FERRIES

As with railways and tramways, ferries may continue to be shown on Soviet maps long after being superseded, even though the replacement bridge or

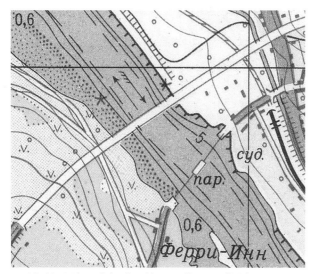
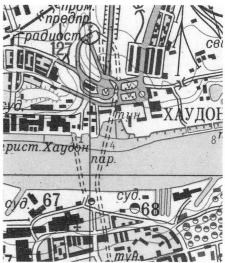

3.100 Erskine Bridge and abandoned ferry on *Glasgow and Paisley*, UK, 1:25,000, 1981.

3.101 Tyne Tunnel and abandoned ferry on *Newcastle upon Tyne*, UK, 1:25,000, 1977.

tunnel appears alongside on the map. Two examples are the ferry (*пар*) adjacent to the Erskine Bridge over the river Clyde, Glasgow (fig. 3.100), and that by the Tyne Tunnel at Jarrow, Newcastle upon Tyne (fig. 3.101). In all, six ferries are shown crossing the river Clyde in the stretch between Glasgow and the Erskine Bridge (only three as operational on contemporaneous OS maps), and six are shown crossing the Tyne, only one of which was by then still operating and shown on OS maps.

ROADS

Road numbers, like place-names, have to be collected from documentary sources, such as maps, atlases, and guidebooks. On Soviet maps of the United States, according to the marginalia (fig. 3.102), federal highways are numbered within a rectangle, with interstates having the prefix T (for transcontinental). State highway numbers are shown within a circle. The translation is "Road numbers: a) Government (transcontinental roads marked with letter T), b) State." A good example is shown in figure 3.103, where I-80 meets US-46 and state route 23 near Paterson, New Jersey. The Soviet research seems to have been well conducted, and few anomalies or errors have so far been detected (but see figure A1.29 for an example).

3.102 Marginalia on *Boston*, 1:25,000, 1979.

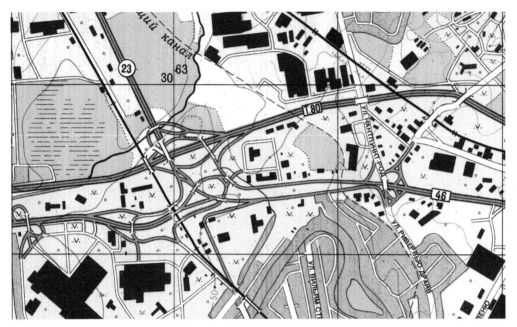

3.103 Highways near Paterson, New Jersey, on *New York*, 1:25,000, 1982.

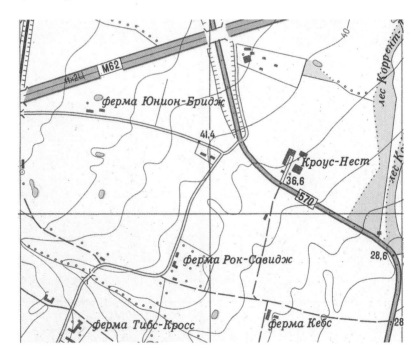

3.104 Road numbering on *St. Helens*, UK, 1:10,000, 1984.

3.105 Road numbering on *Wigan and Ashton in Makerfield*, UK, 1:10,000, 1979.

The situation in Britain was not so straightforward. British roads comprising the national road network are prefixed A, B, or M (first class, second class, motorway, respectively). The Soviet maps are inconsistent in how these classes are portrayed. The general rule is that only rectangles are used, that motorways have an M prefix, and that A and B roads are labeled without a prefix, as seen in the typical example in figure 3.104, showing the junction of M62 and A570 near St. Helens, Lancashire.

However, one highly significant feature appears on Soviet maps, which is the presence of E-numbered road labels. These are European through routes, comprising roads already bearing national numbers but signposted as continuous cross-border international routes. In fact, such numbering has never appeared on British road signs, on any published British maps, or in any public guidebooks or road atlases. Nonetheless, the Soviet maps

PLOTS AND PLANS 115

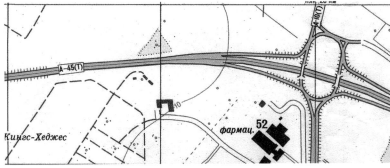

3.106 Road numbering on *Cambridge*, UK, 1:10,000, 1989.

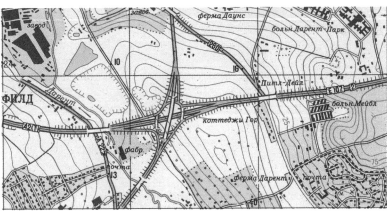

3.107 Road numbering on *London*, 1:25,000, 1985.

do show E numbers on the major highways, alongside the M numbers. The source of the information would presumably be a European-wide official publication and suggests that the cartographers considered the mapping of Britain as part of a pan-European (or wider) endeavor. An example is shown in figure 3.105, where motorway M6 in Lancashire is double-labeled as E33. This has an asphalt surface as indicated by the annotation A, but evidently the dimensions were not determined.

However, anomalies, errors, and inconsistencies abound. In some cases, a hyphen appears between the prefix and the number, sometimes the prefix A or B or the suffix (T) appears (signifying "trunk road"). In some cases, the road number is simply wrong, while in others, road numbers have been confused with junction numbers. Such discrepancies occur extensively across the Soviet city plans of Britain. Figure 3.106, for example, shows a road junction north of Cambridge. Both roads, the A45 and A10, are labeled with prefix, hyphen, and suffix. Figure 3.107, depicting the vicinity

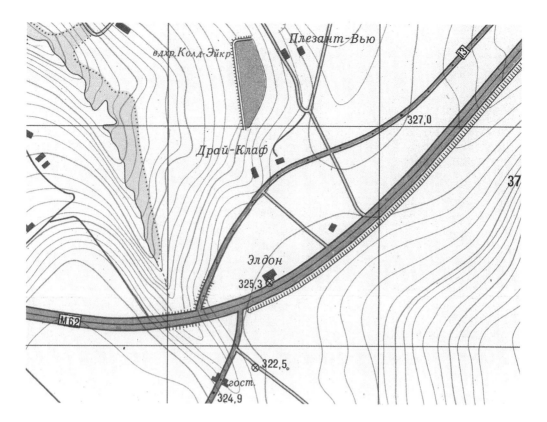

3.108 Errors of road layout and numbering on *Huddersfield*, UK, 1:10,000, 1984.

of the A2/M25 interchange in southeast London, has inconsistencies: the A2 is labeled A2(T), E107, and 2. The A225 is named as such, but the B260 has no prefix.

Several strange errors can be seen in figure 3.108, to the west of Huddersfield, UK. In reality, the A640 runs alongside the M62 to its north and crosses it on an overbridge with no connecting slip roads. Here the road is labeled 13 and appears to join the motorway at a staggered junction. Also, the two minor roads do not join the motorway, which should be shown in a cutting on both sides.

The Birmingham plan has its own unique symbology, as shown in figure 3.109, and its own misconceptions. The key does not explain the significance of the rectangle and circle symbols. The translation reads: "Road numbers (E means part of the European route network)." The numbers 8 and 9 appearing boxed along the M6 motorway in Birmingham (fig. 3.110) are actually the numbers of the nearby junctions. The same error occurs on

3.109 Marginalia on *Birmingham, Wolverhampton, and Walsall*, UK, 1:25,000, 1977.

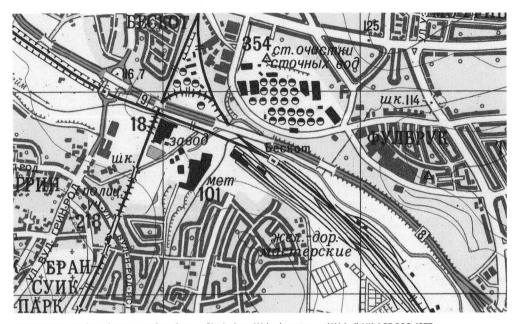

3.110 Junction numbers shown as road numbers on *Birmingham, Wolverhampton, and Walsall*, UK, 1:25,000, 1977.

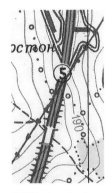

3.111 Road number in circle on *Birmingham, Wolverhampton, and Walsall*, UK, 1:25,000, 1977.

the nearby M5, which also has the motorway number, without the prefix, in a circle as shown in fig. 3.111.

On some plans, a misleading effect occurs where the main-road coloring does not continue through built-up areas, masking the importance of the route. An example is shown at Kingskerswell, north of Torquay, UK (fig. 3.112), where the A380 is the major road to the coast, and northeast of London at Havering-atte-Bower (fig. 3.113), where the B175 is a significant local route. The same can be seen in figures 3.71, 3.72, and 3.73, where the depiction of roads A58, A651, and A652 differ on all three. The effect is particularly strange at Enfield, in north London (fig. 3.114), where the major highway—the A10, Great Cambridge Road—is uncolored, while its less important neighbor to the east, Hertford Road, is colored. The start of the colored stretch of A10 can be seen at the top of the extract.

3.112 Road coloring at Kingskerswell on *Torbay*, UK, 1:10,000, 1976.

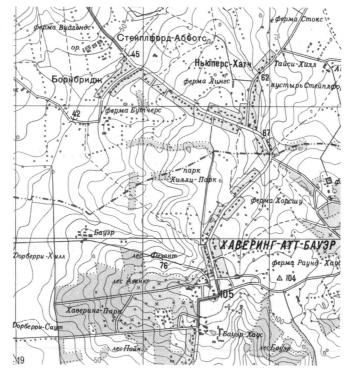

3.113 Road coloring at Havering-atte-Bower on *London*, 1:25,000, 1985.

3.114 Road coloring at Enfield on *London*, 1:25,000, 1985.

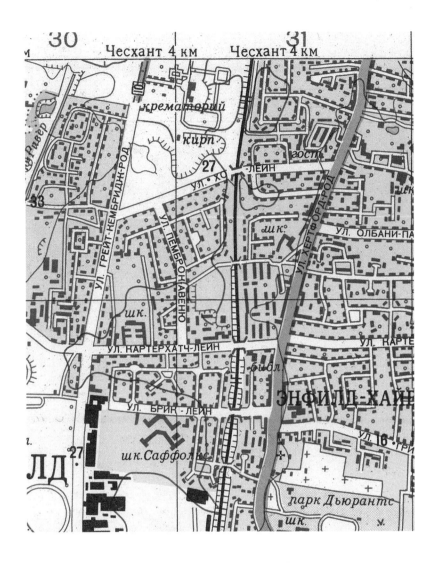

ALL AT SEA

The Soviet maps show considerable detail of marine areas, such as spot depths, bathymetric contours, dredged channels, tidal range, and other details of seas and river estuaries. However, in many cases, this information is at variance from that shown on navigation charts such as those produced by the US Office of Coast Survey and the UK Hydrographic Office (known as Admiralty charts). Neither does Soviet hydrographic information correspond to the data on USGS mapping, even where the adjacent land heights do

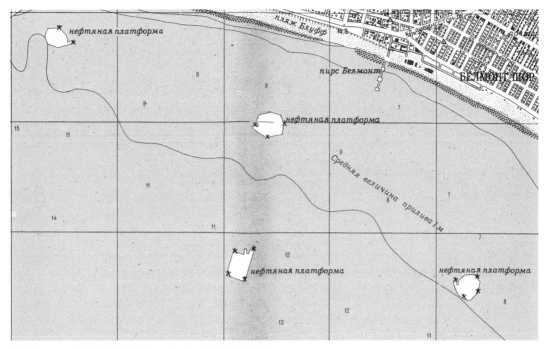

3.115 Astronaut Islands, San Pedro Bay, on *Los Angeles*, 1:25,000, 1976.

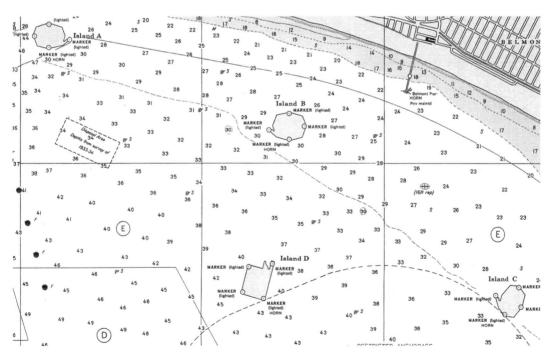

3.116 US Coast & Geodetic Surveys, sheet 5148, *San Pedro Bay*, California, 1:18,000, 1970.

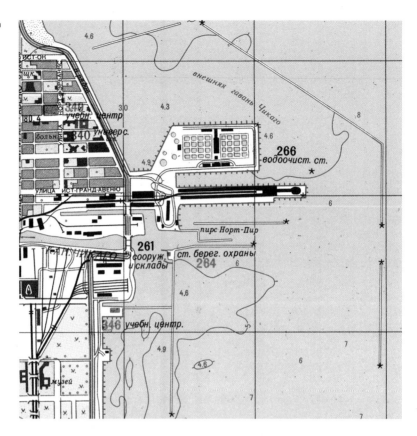

3.117 Hydrographic data on *Chicago*, 1:25,000, 1982.

correspond (British Ordnance Survey maps do not show data beyond the low-water mark.)

A good example of the discrepancy is the depiction of the four artificial islands in San Pedro Bay, California, known as the Astronaut Islands. These were built in 1965 for oil extraction. The Soviet Los Angeles plan—compiled in 1975, printed in 1976 (fig. 3.115)—shows them unnamed but labeled as "Oil Platform"; it shows one as having two lights, one having three lights, and two having four. The surrounding sea is labeled as having a mean tidal range of 1 meter. However, the latest USGS map at the time was dated 1964 and does not show the islands or the tidal range. The US Coast Survey chart, dated 1970 (fig. 3.116), shows the islands named as A, B, C, and D, while the 1974 edition names them as Grissom, White, Chaffee, and Freeman. Neither chart indicates their purpose, and both show all four as having four lights, with no mention of tidal range.

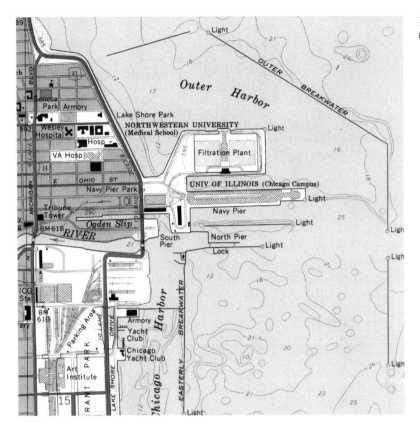

3.118 USGS, *Chicago Loop*, 1:24,000, 1963.

At Chicago Navy Pier (fig. 3.117), the depths such as 4.3 meters in Outer Harbor and 4.6 meters alongside the breakwater disagree with the values of 17 feet (5.2 meters) and 13 feet (4.0 meters), respectively, shown on the then-latest USGS map (fig. 3.118), although the nearby topographic elevations agree (e.g., 592 feet, 180.4 meters). The USGS values and bathymetric contours correspond to the more detailed information shown on the US Coast Survey 1966 chart "Lake Survey 752."

A different story emerges at San Diego, California, where the depiction of channels and seaplane landing lanes in South San Diego Bay on the 1980 Soviet plan (fig. 3.119) is as shown on the 1969 US Coast Survey chart, although the 1970 edition (fig. 3.120) and subsequent charts omit the delineation of the seaplane lanes.

In the case of British coastal waters and estuaries, the situation is quite confusing, and no simple answer can be given to the question of what use

3.119 Seaplane landing lanes on *San Diego*, California, 1:25,000, 1980.

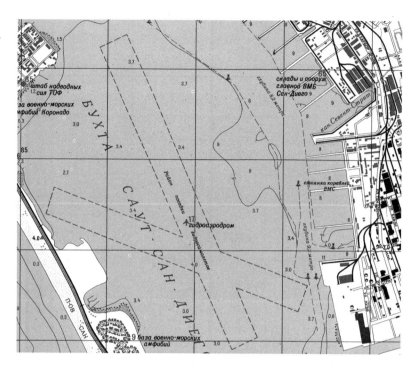

3.120 US Coast & Geodetic Surveys, sheet 5107, *San Diego Bay*, California, 1:20,000, 1970.

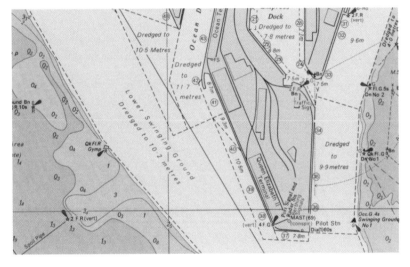

3.121 Hydrographic data on *Southampton*, UK, 1:10,000, 1986.

3.122 UK Hydrographic Office, Chart 2041, 1979.

the Russian cartographers made of Admiralty charts. Certainly, some resemblances and similarities can be seen, but alongside data that is clearly from other sources. As a broad generalization, it seems that some features and annotations were copied, but that depths and channels were generally not, and many anomalies can be seen. On the Southampton plan (fig. 3.121), annotated information such as "dredged to 10.2 meters" and "dredged to 9.9 meters" in the west and east channels, respectively, appear to be copied from the official Admiralty chart number 2041 issued by the UK Hydrographic Office (fig. 3.122). The depiction of the adjacent buildings and railways, however, are different, as are the spot depths in the shallow areas on the left.

The 1985 London plan (fig. 3.123) shows a cable tunnel (*кабельн тун*)

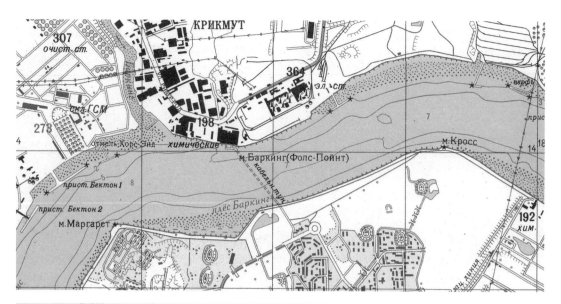

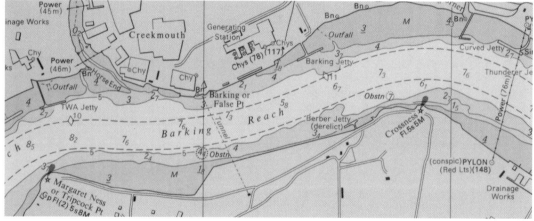

3.123 Underwater cable tunnel at Barking on *London*, 1:25,000, 1985.

3.124 UK Hydrographic Office, Chart 2484, 1978.

under the river Thames, which is shown on the 1978 edition of Admiralty chart number 2484 (fig. 3.124), but not on the previous edition or any Ordnance Survey maps. However, other information, such as the spot depths in the river, was not copied from the Admiralty chart.

On the Glasgow plan, two reaches of the river Clyde, downstream and upstream of the Erskine Bridge, respectively, are both labeled "dredged to 8.0 meters (1963)" and nearby "average height of high tide 3.9 meters." The adjacent Forth & Clyde Canal is annotated with a width of 19 meters and a depth of 2.6 meters. The relevant Admiralty chart—number 2007, dated

1972—shows the downstream reach as "dredged to 27 feet (1963)" and the upstream as "dredged to 26½ feet (1963)"; it has no mention of tidal range and labels the canal as "closed to navigation." In fact, 27 feet is 8.2 meters and 26.5 feet is 8.1 meters. The previous edition of chart number 2007 has the same dredged values but with the date 1957. Similar stories can be told of the Mersey estuary at Liverpool, the Forth at Edinburgh, the English Channel at Dover, and the rivers Tees, Tyne, Medway, and others.

So if not copied from existing charts, where did the information come from? The contents of documents held in The National Archives (TNA) at Kew, London, may suggest an answer. Item ADM 1/28642, "Control of Soviet Vessels in British Territorial Waters 1963–65," comprises bundles of correspondence from which it is clear that frequent visits were made to British ports and territorial waters by a variety of Soviet vessels and that the authorities experienced considerable difficulties in discovering their purpose. Typical is a memorandum from the Ministry of Defence to the Foreign Office dated June 17, 1964, noting that the Soviet Embassy asked for the appropriate authorities to be notified of intended visits by Academy of Sciences research ships *Zarya*, *Akademik Kovalevskij*, *Mikhael Lomondsov*, and scientific research ships *Okeanograf* and *Iceberg*, and requesting permission for visits by the expeditionary ships of the Soviet navy's hydrographic service, *Polyus*, *Zubov*, and *Stvor*. This is followed on August 3 by a memorandum noting:

> From information now available following [the] visit of *Polyus, Zubov* and *Stvor* to Glasgow, it appears that the Soviet Naval Attaché misinformed us when he said that the ships were fully commissioned warships. On the contrary when the Hydrographer of the Navy visited the ships in Glasgow, he was categorically told that they were not warships but Naval auxiliaries. This certainly seems to have been borne out by the fact that most of the people on board were civilian scientists. Perhaps the Soviet Naval Attaché is to be forgiven for misleading us on this point as, in the presence of the Captain-in-Charge, Clyde, he was unable to identify the Soviet Naval medal ribbons worn by the officers of the *Polyus* and by way of explanation let it drop that this was his first job in the navy.

Attached to the memorandum is a report of a visit to the survey ship *Polyus* on July 4 by Rear Admiral E. G. Irving, Hydrographer of the Navy, who comments:

Due to the lack of an interpreter, I was unable to make any serious attempt at trying to find out what the ship was doing and any other technical points. [. . .] many glasses of vodka [. . .] nothing of any special interest [. . .] all the equipment seems quite orthodox [. . .] a number of echo-sounding machines were Kelvin Hughes but I did not see a precision depth recorder. I was not given the opportunity to visit the other two ships which were lying in the same basin. It was explained to me that the ships were naval auxiliaries not "white ensign."

An attached note from the Director of Naval Intelligence dated July 13, 1964, reads:

Despite the Russian explanation that these ships are not "white ensign," we will treat them as naval ships for future visits [. . .] to avoid the uncertainties which happened over the Southampton visit.

There is no explanation of what the difficulties were in Southampton, but the remark indicates that there was unease caused by these visits, and it seems reasonable to conclude that at least some of the information on the maps came from surveys conducted under the cover of visits such as these.

SPRAVKA AND STREET INDEX

Most of the large-scale city plans include three text components: spravka (Справка), a list of important objects, and a street index. The spravka typically comprises a 2,000- to 3,500-word essay describing the city and locality: its physical geography and geology, ethnicity of the citizens, climatic conditions, layout, notable buildings, economic importance, industry, utilities, transport links, and so on. Some of this information could have been derived from generally available sources, but there is much that implies extensive data gathering of local non-cartographic material. A typical example is shown in appendix 3, the translation of the text for the British university city of Cambridge. This makes for fascinating reading. The attention to detail in respect to practical matters such as road construction materials may be expected; more surprising is the attention to cultural details like the decoration of the college gates and the similarity of the lecture halls to ancient castles. The list of important objects (Перечень Важных Объектов) identifies specific buildings and installations that appear numbered on the map.

Table 3.2 shows an analysis of this list for twelve typical plans, six of the United States and six of the British Isles. What is striking is that for the US cities, all the factories depicted on the map have their product identified, and a high percentage (except in Los Angeles) identify the name of the company. By contrast in UK and Ireland, fewer have the product listed and far fewer (except in Oxford) show the company name. This seems to imply that the ability to gather such detail depended on the presence of somebody on the ground and that this may have been more readily accomplished in the United States.

An extract from the list of important objects for Boston, Massachusetts, is shown in figure 3.125, with its translation in figure 3.126; and for Dublin, Ireland, in figure 3.127, translated in figure 3.128. From these it is clear that as much information on production and ownership as possible was collected. The asterisks by the numbers 3 to 10 in Dublin are explained at the end of the list as "unknown production."

TABLE 3.2

USA city	Number of listed objects	Number of which are factories, industrial plants, commercial premises, etc.	Number of which the product or purpose is listed	Number of which the name of the company is identified	Percentage of factories, etc. identified by name
Boston	314	73	73	12	56%
Los Angeles	500	184	184	27	15%
Miami	278	167	167	105	63%
San Diego	90	30	30	29	97%
Seattle	134	52	52	46	88%
Washington DC	252	23	23	18	78%
UK/Irish city					
Birmingham	395	197	182	9	5%
Chatham	70	18	13	1	5%
Dublin	63	15	5	5	33%
London	374	142	142	1	1%
Oxford	41	10	9	8	80%
Southampton	87	19	3	3	16%

61	Завод радиоэлектронный и лаборатория научно-исследовательская	61	Plant electronics and laboratory research
62	Завод радиоэлектронный Рейтсон-Корпорейшен	62	Radio-electronics factory Wrightson Corporation
63	Завод радиоэлектронный Транзитрон-Электроник-Корпорейшен	63	Radio-electronics factory Transitron Electronic-Coproration
64	Завод ракетный	64	Missile factory
65	Завод резинотехнических изделий	65	Rubber products plant
66	Завод резинотехнических изделий	66	Rubber products plant
67	Завод резинотехнических изделий	67	Rubber products plant
68	Завод станкостроительный	68	Machine tool plant
69	Завод станкостроительный Юнайтед-Шу-Машинери-Корпорейшен	69	Machine tool plant United Shoe Machinery Corporation
70	Завод судоремонтный Бетлехем-Стил-Корпорейшен	70	Ship repair factory Bethlehem Steel Corporation
71	Завод судоремонтный военно-морских сил Бостон-Нейвл-Шипярд	71	Navy ship repair yard Boston Naval Shipyard
72	Завод судостроительный Дженерал-Дайнемикс-Корпорейшен	72	Shipbuilding yard General Dynamics Corporation

3.125 List of important objects, *Boston*, 1:25,000, 1979.

3.126 Translation of list of important objects, *Boston*.

1	Аэродром	1	Aerodrome
2	Банк	2	Bank
3	*Группа промышленных предприятий	3 *	Group of industrial enterprises
4	*Группа промышленных предприятий	4 *	Group of industrial enterprises
5	*Группа промышленных предприятий	5 *	Group of industrial enterprises
6	*Группа промышленных предприятий	6 *	Group of industrial enterprises
7	*Группа промышленных предприятий	7 *	Group of industrial enterprises
8	*Группа промышленных предприятий	8 *	Group of industrial enterprises
9	*Группа промышленных предприятий	9 *	Group of industrial enterprises
10	*Группа промышленных предприятий, в том числе завод по производству железнодорожного оборудования	10 *	Group of industrial enterprises, including plant for the production of railway equipment
11	Завод газовый	11	Gas plant
12	Завод газовый	12	Gas plant
13	Завод металлообрабатывающий фирмы Джон-Уиайтакер (Холоууэйр) Лимитед	13	Plant metalworking firm John Whittaker (Alloys) Ltd
14	Завод металлообрабатывающий фирмы Джордж—Милнер-энд-Сонс Лимитед	14	Plant metalworking firm George Milner and Sons Ltd
15	Завод металлообрабатывающий фирмы Смит-энд-Пирсон Лимитед	15	Plant metalworking firm Smith & Pearson Ltd
16	Завод металлообрабатывающий фирмы Хели-Том Лимитед	16	Plant metalworking firm Heli-Tom Ltd
17	Завод электротехнический фирмы Аррелл--Электрикал-Эксессерис Лимитед	17	Plant electrical company Arroll-Electrical Accessories Ltd

unknown production

3.127 List of important objects, *Dublin*, 1:10,000, 1980.

3.128 Translation of list of important objects, *Dublin*.

RESURRECTION

THE STORY OF WHAT HAPPENED TO THE MAPS AFTER THE COLLAPSE of the Soviet Union is in many ways as intriguing as the story of their production. During Soviet times, maps (at all scales of all world locations) were stored in about twenty-five military depots throughout the USSR, where they could be quickly accessed by locally based officers if needed. With the collapse of the USSR, the fate of these maps depended on where they were stored. Those in depots in Belarus, the Russian Federation, and Ukraine remained under Russian control. Gradually, channels of communication—official, semi-official, clandestine, criminal—became established whereby maps were traded with parties in the West in exchange for much-needed hard currency. Few of these transactions have been documented, and for the most part, those involved prefer to keep it clandestine. However, various tales of derring-do have surfaced, such as that noted by Greg Miller [24]:

> A military helicopter was on the ground when Russell Guy arrived at the helipad near Tallinn, Estonia, with a briefcase filled with $250,000 in cash. The place made him uncomfortable. It didn't look like a military base, not exactly, but there were men who looked like soldiers standing around. With guns.
>
> The year was 1989. The Soviet Union was falling apart, and some of its military officers were busy selling off the pieces. By the time Guy arrived at the helipad, most of the goods had already been off-loaded from the chopper and spirited away. The crates he'd come for were all that was left. As he pried the lid off one to inspect the goods, he got a powerful whiff of pine. It was a box inside a box, and the space in between was packed with juniper needles. Guy

figured the guys who packed it were used to handling cargo that had to get past drug-sniffing dogs, but it wasn't drugs he was there for.

Inside the crates were maps, thousands of them.

Map dealers in the United States, such as Guy and others [55], gradually acquired large numbers of paper maps and subsequently sold them to libraries and collectors, and continue to sell digital images through their websites. The Russian government restricted what was made available—for example, maps of the Russian Federation at 1:50,000 and larger were not released.

It was a different story in Latvia. In 1993 Aivars Zvirbulis—a keen orienteer in the town of Cēsis, 100 kilometers east of Riga—had acquired a secondhand printing press in order to print orienteering maps. He heard a rumor that a couple of Russian officers were disposing of quantities of waste paper for pulping. On investigating, he learned of the existence of the Soviet map depot and discovered that its complete contents, over 6,000 tonnes, had been ordered to be destroyed. He was intrigued and negotiated with the officers to buy about 100 tonnes. These were moved by palette load to the yard of his printing house, where, unfortunately, local children set fire to them. Only two or three tonnes were saved, and of the surviving stock, many were training maps or were not of interest. (Zvirbulis was particularly interested in large-scale maps of Baltic countries and western Europe and smaller-scale maps of elsewhere.)

He showed some of the maps at the International Cartographic Conference in 1993, which was held in Cologne, Germany, and attended by some Russian visitors, former KGB officers, who were astonished to see them. They threatened to cause real trouble but were powerless to do so. Zvirbulis went on to establish the Jana Seta map shop and cartographic company in Riga. The Soviet maps were not intended as a core part of the business, and he sold them cheaply. In fact, they proved very popular, and the shop continues to be a magnet for collectors [54].

Another participant at the Cologne conference was David Watt, a British map enthusiast [50] with a personal and professional fascination with all things cartographic, especially those with military connections. He established links with the Riga shop and another in Tallinn, Estonia, and it was not long before British libraries and collectors had started to acquire sets of Soviet maps. In 1996 British map-seller David Archer issued a useful catalog [5], and in 2001 Cambridge University Library published an information sheet "Where to Purchase Soviet Military Mapping" [17], which listed twelve suppliers in North America and in nine countries, including Russia and the Baltic states. In the same year, the British Library displayed the large-scale London plan as part of its major exhibition the *Lie of the Land*.

This was the era before the ready availability of the kind of geo-information freely available today, such as online maps, street views, and satellite images. Commercial mapmakers in Britain wanting to produce and sell maps would have had only three choices for deriving the necessary source information: do their own survey (time-consuming), use out-of-copyright Ordnance Survey maps (at least fifty years out-of-date), or pay license fees to the OS (expensive). The sudden influx of cheap, accurate modern mapping looked like it would change all that.

At the time, the Ordnance Survey was engaged in a long-running dispute with the British motoring organization the AA over alleged copyright infringement (which it finally won with a £20 million out-of-court settlement in March 2001 [2]) and was determined to avoid this potential undermining of its business model. In a statement dated September 10, 1997, the Ordnance Survey declared that the Soviet mapping "is almost entirely an adaptation of Ordnance Survey Crown copyright material. It was produced without the permission of Ordnance Survey and thus it infringes Ordnance Survey's Crown copyright." They demanded that anyone in possession of the maps hand them over and threatened legal action against anyone importing or offering for sale or reproducing any part of the maps [51]. In fact, the statement explicitly refers to small-scale mapping, not the large-scale city plans; and the factual veracity of the statement is debatable in any case. However, the statement effectively buried interest in the maps in the UK for many years.

Internationally, however, the cat was truly out of the bag. The maps proliferated on websites and in collections, public and private, and proved hugely valuable in providing reliable coverage in parts of the world where little or no mapping otherwise existed. Explorers, adventurers, NGOs, and even the military soon discovered them and quickly benefited. Many examples can be shown; the testaments quoted throughout the rest of this chapter are from private communications with the authors.

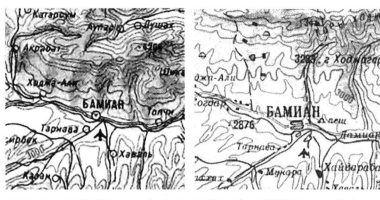

4.1 Extract from 1:1,000,000 topographic sheet I-42, *Bamian*, Afghanistan.

4.2 Extract from 1:500,000 topographic sheet I-42-1, *Bamian*, Afghanistan.

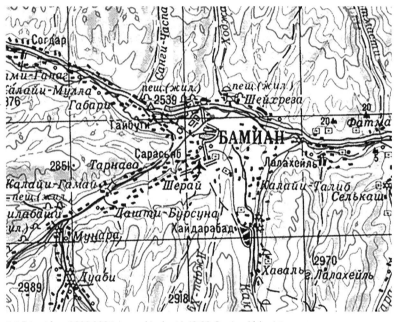

4.3 Extract from 1:200,000 topographic sheet I-42-08, *Bamian*, Afghanistan.

Damon Taylor says:

I was a Lt. Col. in the New Zealand Army and was the Intelligence Officer when the NZ Army went into Bamian in Afghanistan in 2003. We initially relied solely on Soviet maps due to the lack of anything else [see figs. 4.1, 4.2, and 4.3] (although we eventually moved to Allied-produced mapping as the campaign progressed).

Similarly, Thom Kaye recounts:

> I used Russian maps extensively as I mapped Afghanistan for the American National Geospatial-Intelligence Agency prior to the US invasion. We obtained the maps from a British library as we had no other contemporary sources. Although we had used old 1950s Fairchild maps, we had nothing that was as comprehensive as the Russian 1:50,000 or 1:100,000 scale maps. The one thing that really struck me was the mountain passes. The Russian maps actually had listed on the map face the dates at which the passes were clear for travel. This tells me that almost persistent surveillance was done to capture this information. Moreover water sources like wells and springs were carefully extracted due to the arid environment and the obvious need for a freshwater source. Satellite imagery can only grant you so much information, so these maps were invaluable.

Craig Jolley, an AAAS Science & Technology Policy Fellow working with the US Agency for International Development to support water resource management in Armenia's Ararat Valley, told us:

> One challenge that we ran into is that it's very hard to find historical information on the location or condition of artesian wells in Armenia. Having now looked at the Soviet topo maps, we find that artesian wells are clearly labeled on the maps, often with flow rates. Since we're interested in updating the Ararat aquifer data, historical information on flow rates could be useful for determining where depletion has been severe and convincing people that action really is needed. I think these maps could be valuable for looking at changes in environmental conditions, types of land use, aquifer levels, etc.

Stuart Pask, who works on the geophysical and geological surveying and mapping that is carried out at the beginning of the oil exploration process, explains:

> Back in the mid-1990s, we used 1:50,000 Russian mapping as the starting point for our fieldwork. It was in the early days of GIS systems, and we typically had the map scanned and would use them as a transparency overlaid on satellite imagery. From this, we could plan surveys and get a feel for land use etc. and then plan scout trips out to ground-truth everything

before we started the survey work in earnest. I used the same techniques again in India in the late 1990s and more recently in Nepal and Cameroon.

As GIS and GPS technology evolved, we would have a laptop in the back of a vehicle running ArcGIS with the Russian maps overlaid on satellite imagery with a real-time position of the car plotted on the screen. It was a dream solution for scouting new work areas.

Having everything loaded onto a GIS system also got round the challenges some countries have with mapping and security. In India, they would confiscate paper maps and satellite imagery if you arrived at the airport with maps in a tube, but were blissfully unaware of the mapping capabilities that we had installed on a laptop!

Colonel Desmond Travers (retired) of the Irish Army served on several UN-appointed missions in Israel and Lebanon in the 1980s, as he relates:

My duties included the investigation of incidents close to Israel's borders with Lebanon, and I needed an understanding of the historical context in which these occurred. I relied largely on maps produced by the British Mandate—for Israel and Palestine—and the French Administration for Lebanon and Syria. I observed that on these maps man-made topography had over time been politicized. Habitations and even ancient heritage sites had been erased from the landscape and place-names changed to bring them into line with the new, often religious, austerities and so on.

Later, in retirement, I served as military analyst for a human rights organization during the Hezbollah-Israeli war in South Lebanon in 2006 and as a member of the UN-appointed fact-finding mission into the Israeli conflict in Gaza "Operation Cast Lead" in 2009. In both of these incidents, and on subsequent visits, I was now able to use Soviet maps. They were significantly better in the representation of topography, being derived from satellite imagery and in color. But what I again observed was the politicization of topography. The explanation may be that the Soviet maps were "military" rather than administrative in purpose. They were more interested in "the going"—that is, road, rail, and air networks—than the more social features. They were also interested in places of production such as water mills, power stations, manufacturing and repair facilities. They had no interest in the representation of heritage or historical sites even though the "Fertile Crescent"—the coastal Levant—was the cradle of civilization, or at least so for those [who] were to migrate through there into Europe and Asia. For example, such significant archaeological features as a "Canaanite Temple" would not be represented. Ironically this was to bring their maps into line with the more recent maps of the region, which were concerned in the representation of the newer Shi'a Islamic presence (South Lebanon) and the emerging Jewish tradition (in Israel).

Travers adds that on the Soviet maps of his native Ireland, long-abandoned rural water mills of no importance are shown whereas many ecclesiastical buildings or ruins, far more significant in Ireland's cultural landscape, are ignored (see fig. 4.4).

The legacy of the Soviet mapping project lived on in other ways too. Upon its independence, Latvia established a national mapping agency, the State Land Service, whose specification of the symbology for topographic maps [10] was derived from the established Soviet specifications, having the familiar annotation of the dimensions and values of characteristics of roads, rivers, bridges, forests, and so on (figs. 4.5 and 4.6). The Latvians also used the Soviet maps as a source of information for place-names (fig. 4.7).

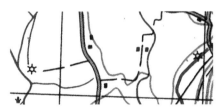

4.4 Extract from 1:100,000 topographic sheet N-29-119, *Carlow*, showing two water mills (the star-shaped symbol) in rural Ireland.

4.5 and 4.6 Extracts from Latvian specifications for 1:50,000 topographic maps published in 2000 by the State Land Service of the Republic of Latvia showing the typical Soviet-style annotation of characteristics of roads and forests.

Automaģistrāle ar melno segumu 7 - brauktuves platums, 10 - ceļa platums, A - seguma materiāls *Black surface motor highway* *7 - surface width* *10 - overall width* *A - surface material*	7(10) A
Skujkoku mežs *Coniferous forest*	egle
Lapu koku mežs *Foliage forest*	bērzs
Jauktu koku mežs *Mixed forest*	priede kļava

RESURRECTION

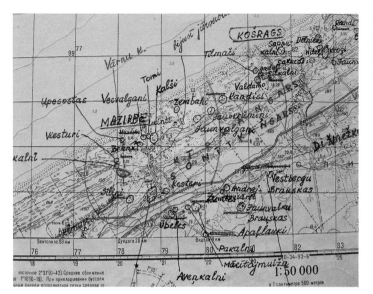

4.7 Extract of a Soviet 1:50,000 map (O-34-081-4) covering part of Latvia, with place-names annotated by hand and in Latvian, in use at the Latvian Geospatial Information Agency (LGIA).

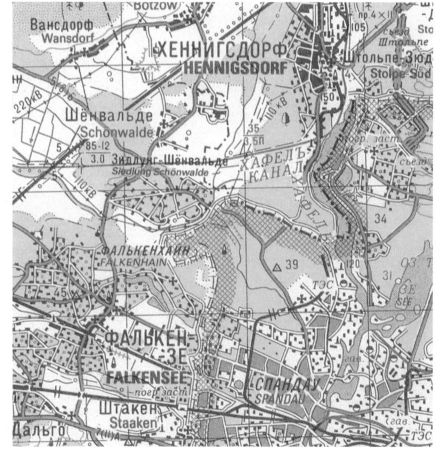

4.8 Extract from a bilingual edition of 1:200,000 sheet N-33-32, *Berlin (West)*, produced and published in 1986 by Ministerium für Nationale Verteidigung Militärtopographicgraphischer Dienst (DDR Ministry of National Defense Military Topographic Service).

However, in 2006 responsibility for cartography passed to a new organization, the Latvian Geospatial Information Agency (LGIA), and subsequently the specifications have been simplified and the annotation dropped (it is significant that the reason given is the amount of effort that would be required to collect and maintain the necessary information).

The reunification of Germany and the consequent reconciliation process resulted in the opening up of doors and books. The MTD (East German Military Topographic Service) had operated in close conjunction with the Ministry for State Security and produced military maps to the standard Warsaw Pact Coordinate System 42 specification, just as the Soviet maps but with Roman script rather than Cyrillic (fig 4.8).They also produced a version known as AV: "Ausgabe für die Volkswirtschaft" (Edition for the National Economy), which omitted geodetic grids and much other "sensitive" information and introduced distortion in direction and scale.

Following reunification, the MTD was absorbed into the MGD, the Military Geographic Service of the West German Bundeswehr, in 1990, and gradually secrets became unlocked as records were opened up. Eventually two remarkable publications appeared that revealed the practices and products of the DDR (which, in effect, corresponded to those of its Warsaw Pact comrades). The first, *State Security and Mapping in the German Democratic Republic: Map Falsification as a Consequence of Excessive Secrecy?* [32], provides vivid evidence of the culture of obsessive secrecy and the often ludicrous extremes to which it was pursued. Not only were huge resources poured into creating alternative maps having distortions and omissions; the falsified mapping had dire consequences for the national economy. Planning and construction of roads and industrial sites became more difficult as companies reverted to using prewar mapping. The other, *Militärisches Geowesen der DDR von den Anfängen bis zur Wiedervereinigung* (East German military-topographic service from inception to unification) [35] is a historical account of the organization and its products, copiously illustrated with examples including military maps, special-purpose maps, AV edition maps, and training maps (the latter being mirror images of standard maps).

The relationship between Finland and the USSR during and after the Cold War period is also interesting. One person closely involved was Erkki-Sakari

Harju, former technical director of the Publications Division of the National Land Survey of Finland, now retired. He recalls:

> I started my work at the Publications Division in 1969. At that time Finland was not allowed to produce maps of the area lost to USSR after [the] Second World War. The reason was that a Soviet-British Control Commission was established to monitor Finland's adherence to the Paris peace agreement. The Soviet members of the Commission ordered that Finland had to deliver all maps and aerial photos, geodetic data, map originals etc. to the Soviet Union from the area which was now under Soviet control (although this was not in the peace agreement). Finland was not allowed to show those areas on Finnish maps, except at the scale of 1:500,000 or smaller. The Finnish Ministry of Foreign Affairs canceled this statute in 1989, at which point everything changed. Until then, the Russian part on our maps was white paper, now it was no longer a forbidden area. We started negotiations in Moscow with the company Sojuzkarta to obtain map data for our 1:200,000-scale road maps, which covered a part of Karelian isthmus and East Karelia. After Moscow we continued the talks in Leningrad with the state enterprise Aerogeodezija [North West Aerogeodetic Institute]. As a result we got the necessary additional map data for our road maps. In this connection we noticed that also the earlier secret topographic maps were available. The Finnish military was particularly interested in them and we started buying the maps. During these talks we noticed that Finland was totally covered by Russian topographic maps in scales of 1:50,000, 1:100,000, 1:200,000, and 1:500,000. The source data for these maps had come from the Finnish topographic maps, which were free to everybody to buy. The best place to buy new 1:20,000-scale topographic maps and all other Finnish maps was the map shop of the National Land Survey. The topographic mapping of Finland was completed at 1:20,000 scale in 1975, and all these maps were bought by the Soviet Union (and by Great Britain for NATO). We used to say that when the Volga of the Russian embassy was round the corner, the Rover of the British embassy arrived to the door of the map shop. We produced also 1:50,000-scale maps for the military (also [available] to buy) and these maps were bought by Russia, too. During one of my visits to Sojuzkarta, I was offered a digital map of

Finland at 1:50,000 scale, price one million Finnish marks. I asked [for] some samples of this map data in raster form, and it was totally clear that the data was a direct copy of Finnish maps. Only the map symbolism was modified to follow Russian standards.

In 1995, I left the Publications Division and went to a private sector cartographic company Geodata. We produced (among other things) digital terrain models for the telecommunication companies. These models were needed, for example, by Turkey, Algeria, and some other African countries. To create these models, we bought Russian maps and digital data from the company Priroda in Moscow.

That the West knew of the existence of Soviet mapping is not in doubt, as the US Army Technical Manual Manuals of 1957, 1958, and 1963 testify [4, 9, 3]. However, the extent to which the West seriously underestimated the global scope and scale of Soviet mapping (even as late as 1992, after the collapse of the USSR) is illustrated by this extract from a publication by the Geographic Research Branch of the British Military Survey [1]:

> One of the benefits resulting from the break-up of the Soviet Union has been the availability of the official USSR topographic mapping, which not only covers that country but also certain countries in other parts of the world, such as South America, Africa and Asia. The maps are at scales 1:25,000, 1:50,000, 1:100,000, 1:200,000, 1:500,000 and 1:1 million, but the 1:50,000 and 1:200,000 are the most readily available; and incorporate a wealth of detail not previously shown on maps of the former USSR. Examples of the features shown are heights of railway embankments, constructional details of bridges, and the width and surface material of roads, all of which are almost unique in published official topographic mapping. Curiously, airfields are rarely depicted.

It is interesting that neither the city plan series nor the topos of Europe are mentioned.

The secrecy and sensitivity surrounding Soviet mapping continued long after the collapse of the USSR. Although large numbers of topographic maps and city plans of most of the world did gradually seep into the public domain, large-scale mapping of the territory of the Russian Federation remained under tight security. In 2010 a Russian citizen, Gennady Sipachev, was jailed for four years for handing over maps, classified as "state secrets," to the United States through the Internet. "The maps can be used to make the targeting of US cruise missiles against Russian targets more accurate," Russia's security service officials said [8]. Two years later, Colonel Vladimir Lazar was sentenced to twelve years in a high-security prison for handing a computer disk containing 7,000 scanned images of printed maps to a US Intelligence agent in Belarus. Lazar had been serving in the "military-technical department of the general headquarters of the high command," the FSB (Federal Security Service) said in a statement [6].

In recent years, interest in the Soviet mapping project has been particularly strong in Sweden, a country that, because of its strategic location on the Baltic coast, has long felt uneasy under the Russian gaze (as the story quoted in chapter 3 shows). Swedish military researchers have published detailed accounts of the history of the Swedish-Russian relationship and the role of the military maps [42, 43].

The time, effort, and resources required to produce each Soviet map or plan raise obvious questions about the use for which these tremendous repositories of geospatial intelligence were actually intended. Why go to such effort under the specter of nuclear Armageddon, after which the resemblance of any urban or rural landscape to its cartographic depiction would be minimal?

If we take the listing of important objects on the Soviet plans (e.g., the highlighting of military, communications, governmental, administrative, and military-industrial sites) together with the various detailed annotations (such as the carrying capacity of bridges and the density of tree cover), this information would certainly be very valuable in conducting a military campaign. Together with this detailed depiction of information, the level of geodetic accuracy, the impeccable organization of sheets at several scales, and the adoption of a conformal projection to preserve bearings (e.g., to facilitate targeting by artillery) would also suggest their role in combat operations.

But as we have shown, the Soviet maps are not primarily concerned with the depiction of enemy military installations, and perhaps other, more secret, maps and plans were made for targeting military sites (such as missile silos) in a counterforce strike against NATO. Indeed, the omission of crucial targets and the presence of such a rich amount of detail in the portrayal of infrastructure (including, e.g., disused railways) might suggest that the Soviet maps were intended to support civil administration after a successful coup.

As products of a military cartographic tradition, however, Soviet maps aim to compile and present the best geospatial intelligence in the clearest way, regardless of their eventual use. The question of whether they were "invasion plans" to facilitate hostile aggression or simply peaceful preparations for the inevitable day when the entire world is communist will, no doubt, be hotly debated for years to come. Either way, they offer unsurpassed inventories of topographic information.

The inheritance of such a colossal topographic mapping program therefore presents a number of pertinent questions that are hard to ignore. It is intriguing that large-scale global mapping evidently continued after the breakup of the USSR. A 1:25,000 plan of Falmouth, a naval port on Britain's south coast, for example, was produced in 1997. It is not marked "Secret" but instead "copyright VTU." Similarly marked is a 2003 plan of Vancouver, Canada.

Stalin's decision to invest in a world military-mapping program to depict terrain and resources in detail around the globe has led to an unparalleled legacy of geographical knowledge and geopolitical potential.

ACKNOWLEDGMENTS

WITH SPECIAL THANKS TO THE FOLLOWING, EACH OF WHOM IN their own way made this book possible, sometimes more than they will ever know: David Archer, Philip Aris, Charles Aylmer, Inta Baranovska, Aivars Beldavs, Elliot Carter, John Cruickshank, the Cunningham family, Alastair Davies, Samuel Fanous, Erkki-Sakari Harju, Craig Jolley, Thom Kaye, Alison Kent, Nick Millea, Greg Miller, Stuart Pask, Tony Swarbrick, Anne Taylor, Damon Taylor, Sarah Tinker, Desmond Travers, David Watt, John Winterbottom, Gerry Zierler.

Particular thanks are due to Martin Davis and John Hills for their help in compiling some of the information and illustrations in this book.

APPENDIX 1:
MAP EXTRACTS

THESE EXTRACTS HAVE BEEN SELECTED TO SHOWCASE THE GEOgraphic scope and stylistic variety of the maps and to illustrate the points raised in earlier chapters.

> Figures A1.1–A1.36 are city plans in alphabetic order.
> Figures A1.37–A1.56 are standard series topographic maps in order of increasing scale.
> Figures A1.57 and A1.58 are special series maps.

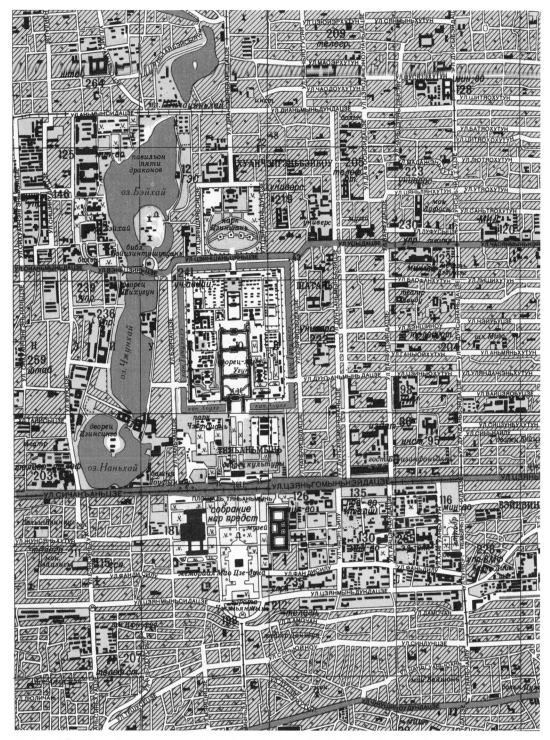

A1.1 *Beijing*, 1:25,000, compiled 1978, printed 1987 (sheet 2 of 2), showing the Forbidden City and Tiananmen Square. Crosshatching over the built-up area indicates that the outline of individual buildings is unknown. Object 181 (*lower center left*) is labeled in the index as "facility for testing missile engines," although this building is the Great Hall of the People.

A1.2 *Berlin*, 1:25,000, compiled 1979, printed 1983 (sheet 2 of 4), showing the international boundary along the Berlin Wall.

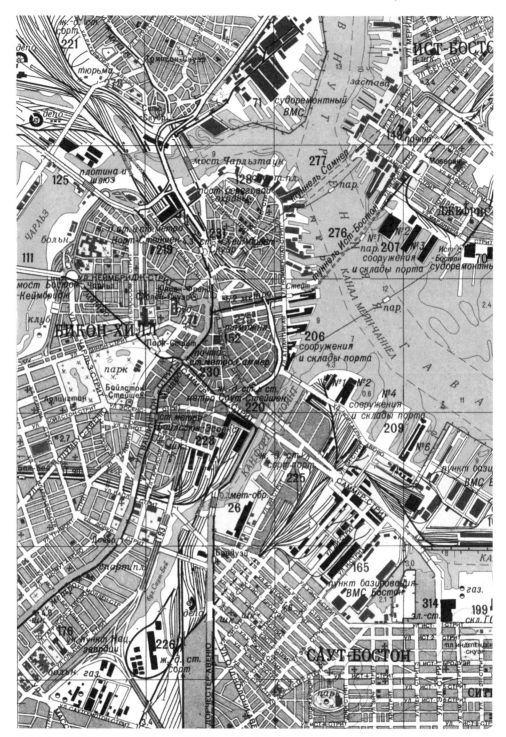

A1.3 *Boston*, 1:25,000, compiled 1977, printed 1979 (sheet 2 of 4). The ferry across the harbor (south of the road and rail tunnels) is not shown on USGS maps, but does appear in local tourist guides. Note the US Navy shipyard (*center top*) with boundary fence, indexed as object 71.

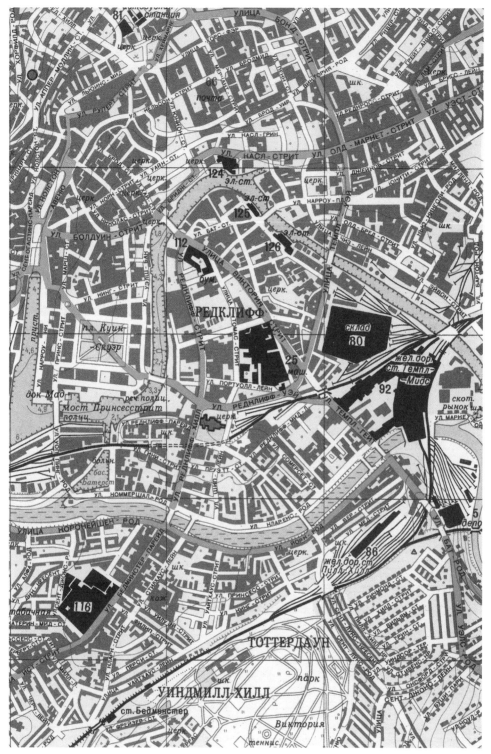

A1.4 *Bristol*, UK, 1:10,000, compiled 1971, printed 1972 (sheet 3 of 4). The brown dashes in the river Avon (*lower center*) indicate silty tidal shallows; the blue marks along the dockside and channel to the north indicate wharfage.

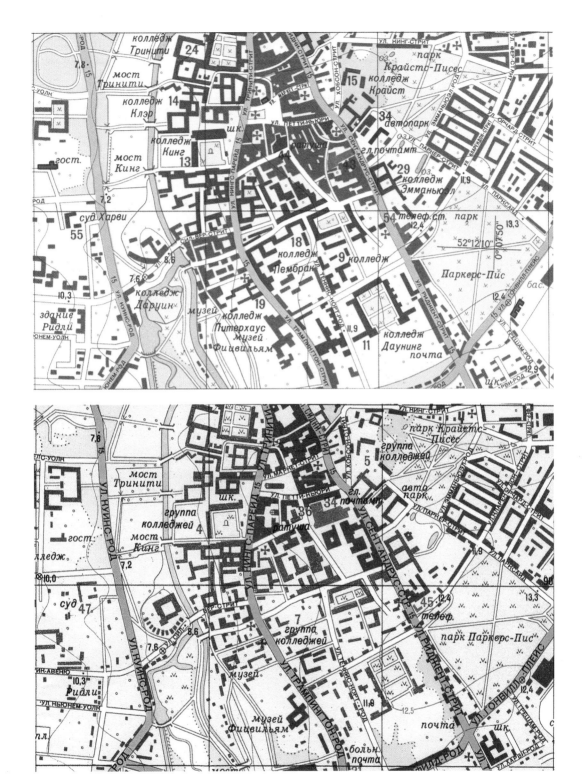

A1.5A and A1.5B *Cambridge*, UK, 1:10,000, compiled 1973, printed 1977 (*above*); 1:10,000, compiled 1986, printed 1989 (*below*).
On the earlier plan, colleges are named individually; on the later plan, they are simply labeled "group of colleges."

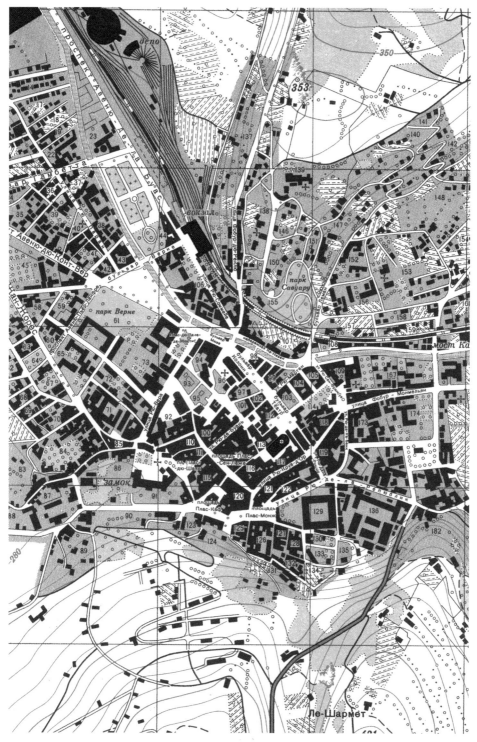

A1.6 *Chambéry*, France, 1:10,000, compiled 1954, printed 1954. An early plan in the pre-1970s style. The map has a brief spravka and a street index but no list of important objects. Each block of land is numbered sequentially. Why this relatively small city in rural France was selected for mapping so early in the project is something of a mystery.

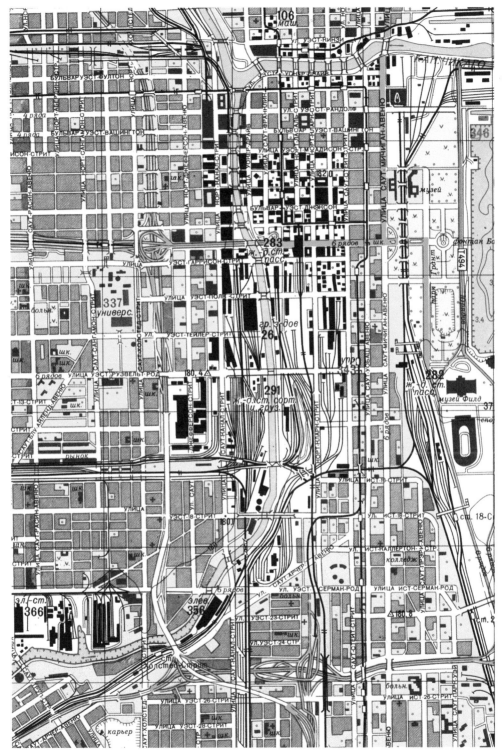

A1.7 *Chicago*, 1:25,000, compiled 1972, printed 1982 (sheet 4 of 7), showing the distinction between industrial blocks in black-and-white and urban blocks, a differentiation not apparent on USGS maps or local city maps.

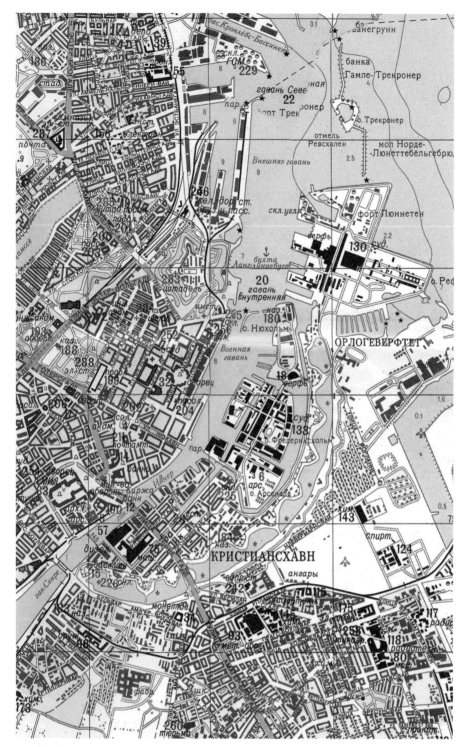

A1.8 *Copenhagen*, 1:25,000, compiled 1982, printed 1985 (sheet 1 of 2). The outline airplane symbol indicating "landing strip" and the label "ангары" (hangar) at the lower right are evidently anachronistic as the airfield (Kløvermarken) was closed in the 1920s and the site used as sports fields.

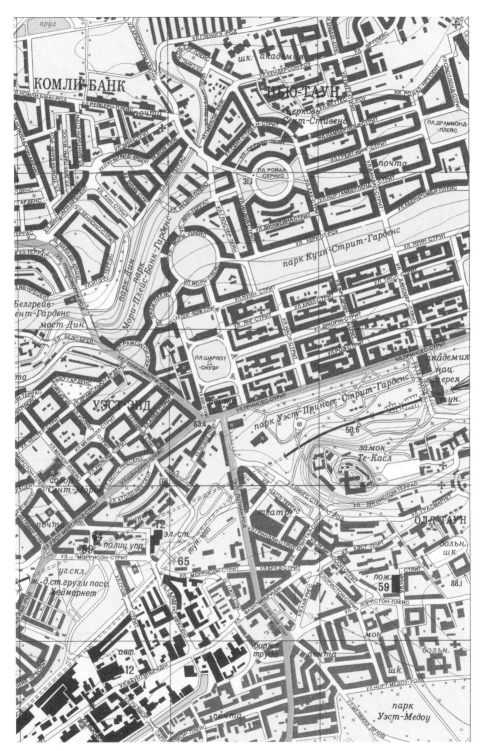

A1.9 *Edinburgh*, 1:10,000, compiled 1980, printed 1983 (sheet 1 of 3, although sheet 3 consists solely of spravka and index). The historic fortress of Edinburgh Castle (*center right*), labeled "Те-Касл" (phonetically, "the castle"), is still partly occupied by the army (albeit largely in a ceremonial role) but is not, as might have been expected, listed as a military object.

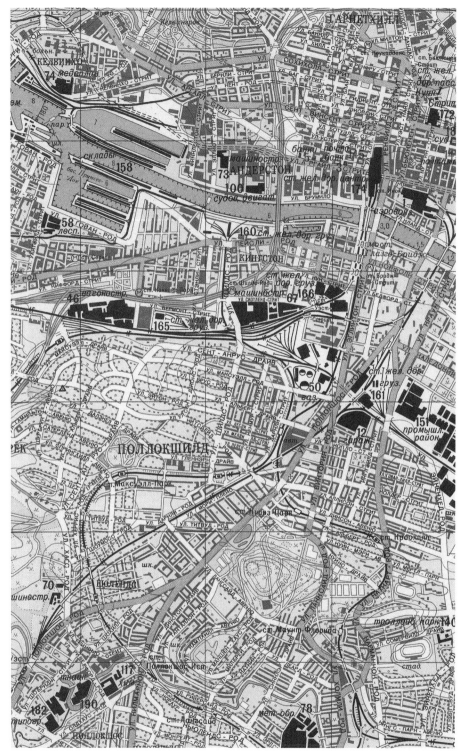

A1.10 *Glasgow and Paisley*, UK, 1:25,000, compiled 1975, printed 1981 (sheet 1 of 2). St. Enoch's station (*upper right, north of the river*) was closed in 1966 and demolished in 1977.

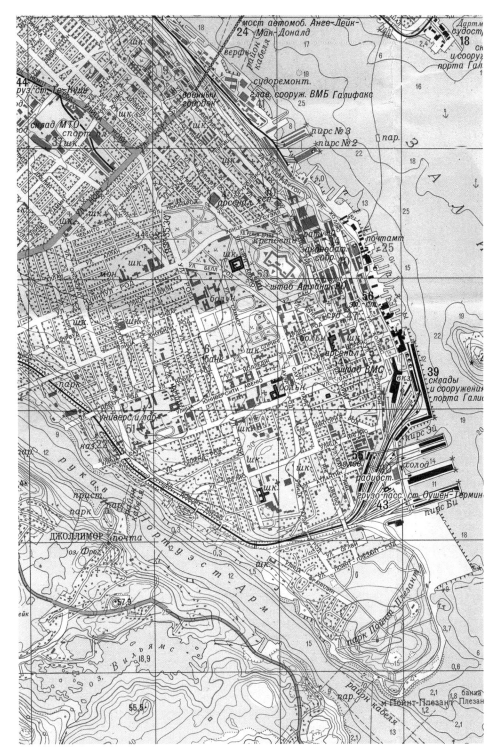

A1.11 *Halifax*, Nova Scotia, 1:25,000, compiled 1973, printed 1974. Object 53 (*center*), Halifax Citadel National Historic Site, is listed as "Atlantic Headquarters Military District (the wartime control point)," although its military significance ended after World War II and it became a national historic site, a tourist attraction rather than a military establishment, in 1956.

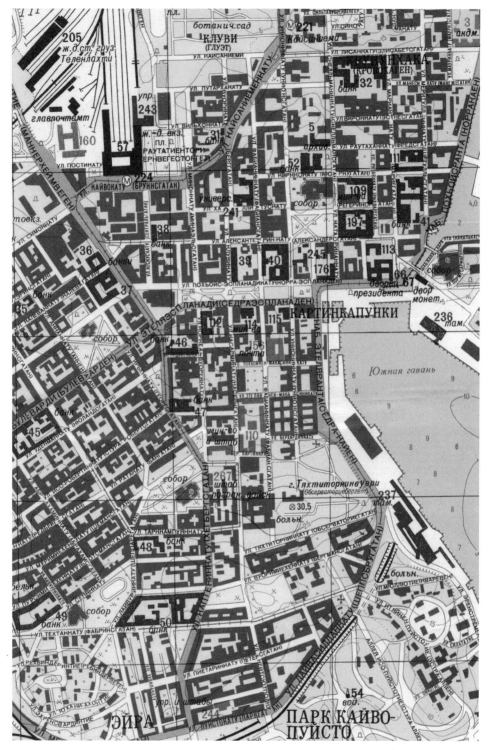

A1.12 *Helsinki*, 1:10,000, compiled 1984, printed 1989 (sheet 5 of 6). The index and spravka were issued as a separate booklet.

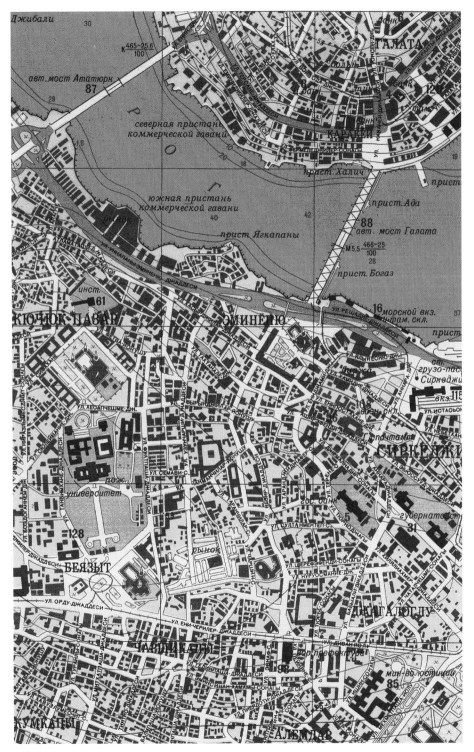

A1.13 *Istanbul*, 1:10,000, compiled 1978, printed 1987 (sheet 4 of 4). The cross-hatching on the eastern bridge (*upper right*, the Galata Bridge over the Golden Horn) indicates a metal bridge (this bridge was replaced in 1992 by the present structure).

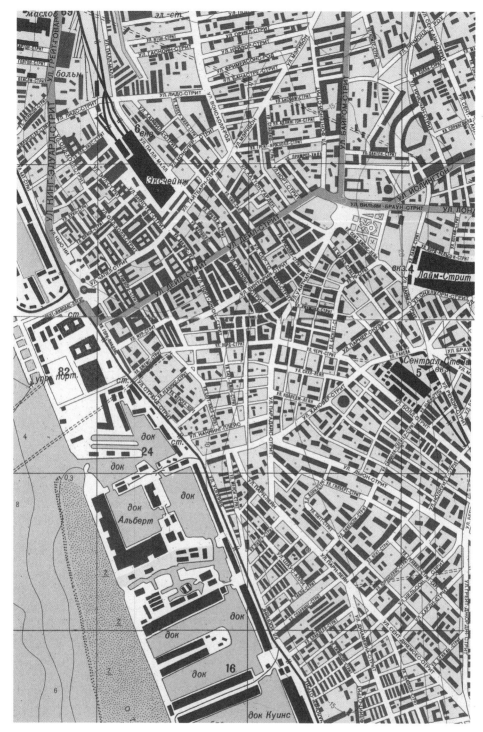

A1.14 *Liverpool*, UK, 1:10,000, compiled 1968, printed 1974 (sheets 1 and 3 of 4). The northwest and southwest sheets of the four-sheet set of Liverpool meet without overlap at the city center. This compilation extract shows a railway running northwest-southeast alongside the docks. This is the Liverpool Overhead railway, which was closed in 1956 and dismantled in 1957, eleven years before the compilation date of the map.

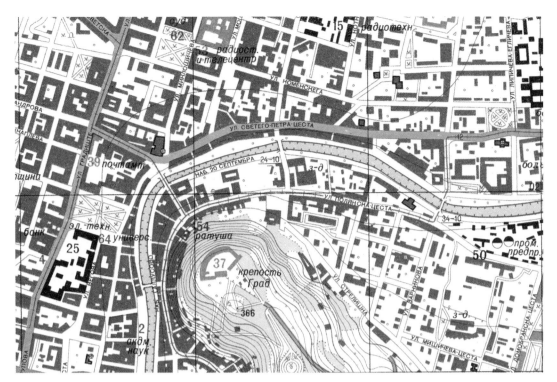

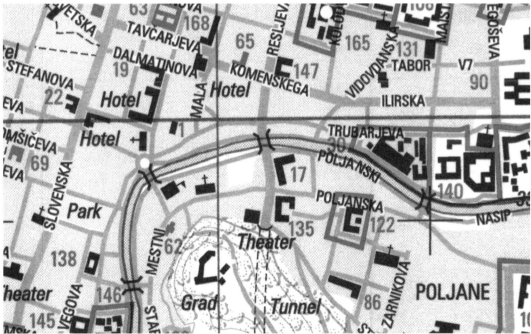

A1.15A and A1.15B *Ljubljana*, Slovenia, 1:10,000, compiled 1976, printed 1980 (*above*) and US National Imagery and Mapping Agency, 1:25,000, 1997 (*below*). Although these plans are of different scales and dates, the contrast in styles and in the levels of detail depicted of the two mapping agencies are clearly seen. The American map is not secret and would have been intended for general use by troops, rather than as a comprehensive repository of geo-intelligence.

A1.16 *London*, 1:25,000, compiled 1980, printed 1985 (sheet 1 of 4). This plan is a masterpiece in portraying a vast amount of detail in a densely packed depiction of the central area.

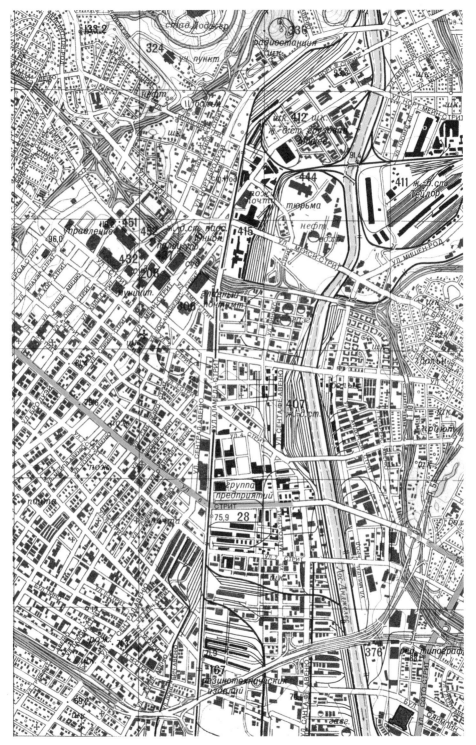

A1.17 *Los Angeles*, 1:25,000, compiled 1975, printed 1976 (sheet 5 of 12). The twelve-sheet of Greater Los Angeles covers more territory than any other known city plan, stretching a distance of almost sixty miles from the San Fernando Hills in the northwest to Santa Ana in the southeast.

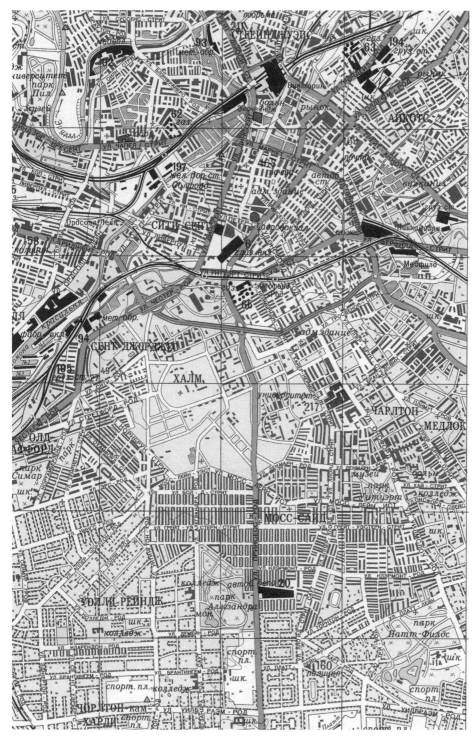

A1.18 *Manchester, Bolton, Stockport, and Oldham*, UK, 1:25,000, compiled 1972, printed 1975 (sheet 3 of 4). The Manchester plan adjoins the Warrington sheet, which together with the St. Helens and Liverpool sheets provides continuous large-scale coverage over the fifty miles from the Irish Sea to the English watershed on the summit of the Pennine Hills.

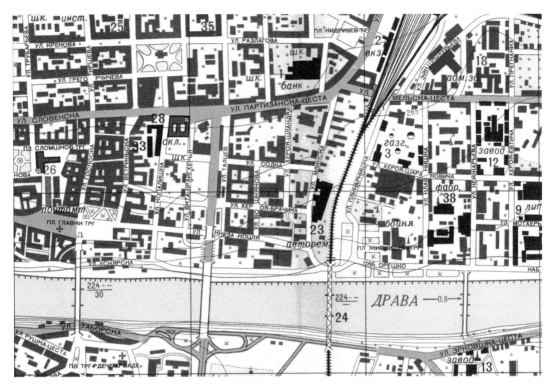

A1.19A and A1.19B *Maribor*, Slovenia, 1:10,000, compiled 1972, printed 1975 (*above*), and US National Imagery and Mapping Agency, 1:20,000, 1993 (*below*). See figure A1.15 for comments about the two mapping styles.

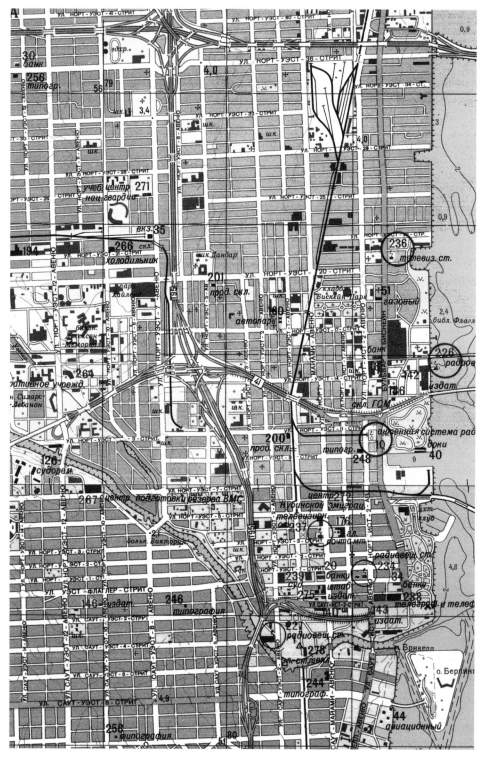

A1.20 *Miami*, 1:25,000, compiled 1981, printed 1984 (sheets 1 and 2 of 2). This compilation shows the north and south sheets where they abut without overlap.

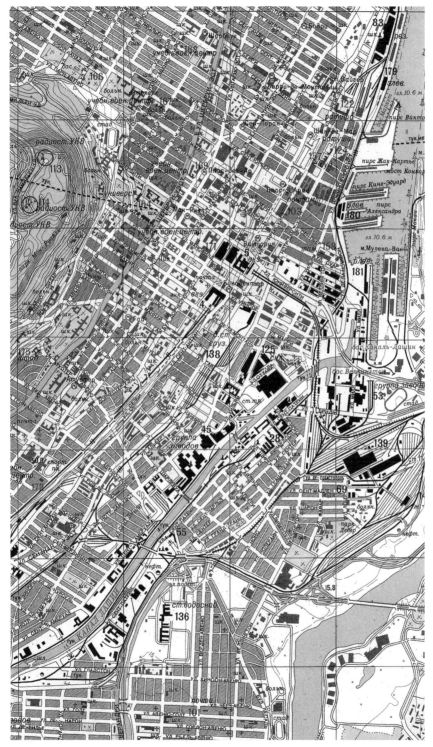

A1.21 *Montreal*, 1:25,000, compiled 1981, printed 1986 (sheet 2 of 2). High-rise blocks (e.g., west of the main road) are shaded light brown; blocks of generally low-rise buildings (e.g., to the east of the road) are shaded gray.

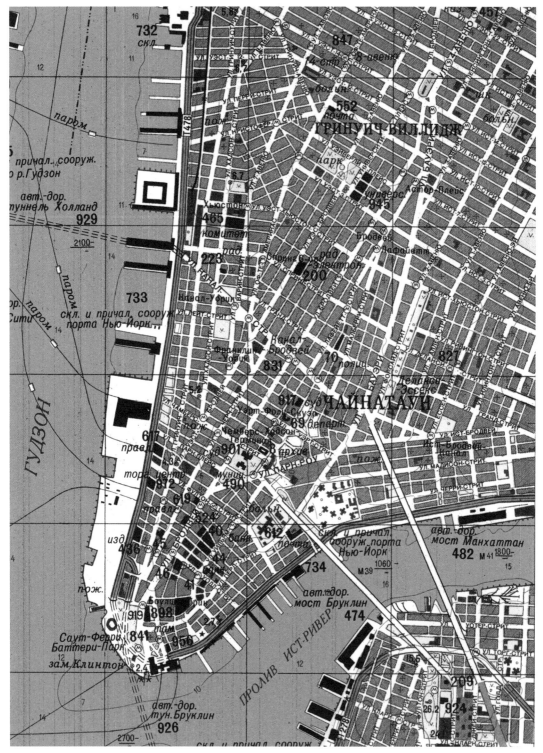

A1.22 *New York*, 1:25,000, compiled 1979, printed 1982 (sheet 6 of 8). Unlike the London plan, the footprints of individual buildings are not shown; the city blocks are generalized and classed as high-rise.

A1.23 *Newcastle upon Tyne, Gateshead, South Shields, and Tynemouth*, UK, 1:25,000, compiled 1974, printed 1977. The main highway north, the A1 (*upper left*) is labeled E31, a designation not used in Britain.

A1.24 *Oxford*, UK, 1:10,000, compiled 1972, printed 1973. Object 39 (*center right*) is labeled "university." In fact, many of the buildings shown (but not identified) are colleges of Oxford University, this one being University College.

A1.25 *Portland*, Maine, 1:10,000, compilation date not stated, printed 1972. One of the few known examples of a 1:10,000 plan of an American city.

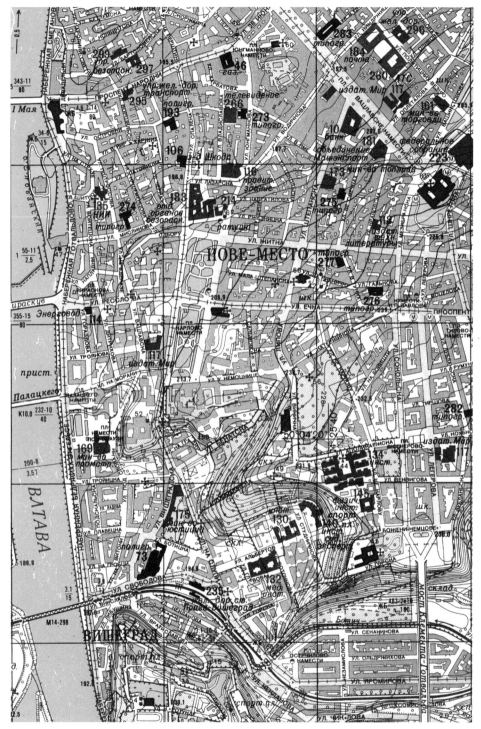

A1.26 *Prague*, 1:10,000, compilation date not stated, printed 1980 (sheet 5 of 9) An example of the wealth of additional information portrayed for cities in Warsaw Pact countries. The contour lines are at 2-meter intervals; the upper and lower water levels of the river are annotated, as are the length, width, and depth of the lock channel at the southern tip of the island (*upper left*).

A1.27 *Raleigh*, North Carolina, 1:25,000, compiled 1976, printed 1980. Most city blocks are shown generalized as low-rise; others have the footprints of individual buildings. The section of the plan to the east of this extract shows housing and highway developments later than the then-current USGS maps.

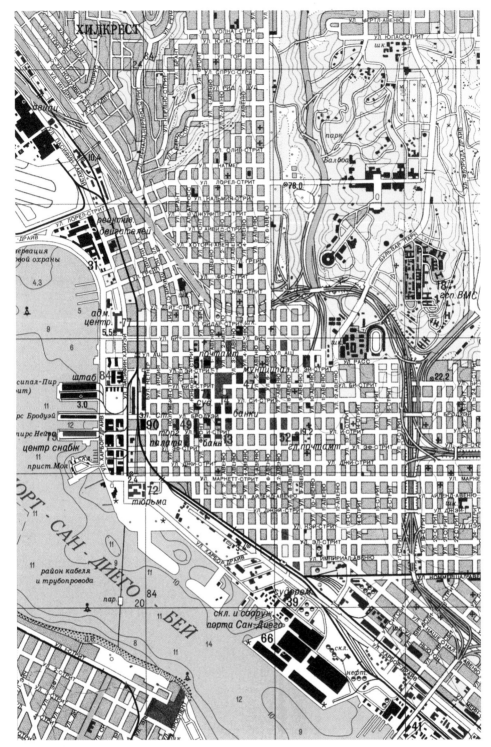

A1.28 *San Diego*, California, 1:25,000, compiled 1975, printed 1980 (sheet 1 of 4). This four-sheet set extends over the Mexican border to Tijuana, which is shown in the same style and the same level of detail (although USGS maps show only rudimentary information south of the border).

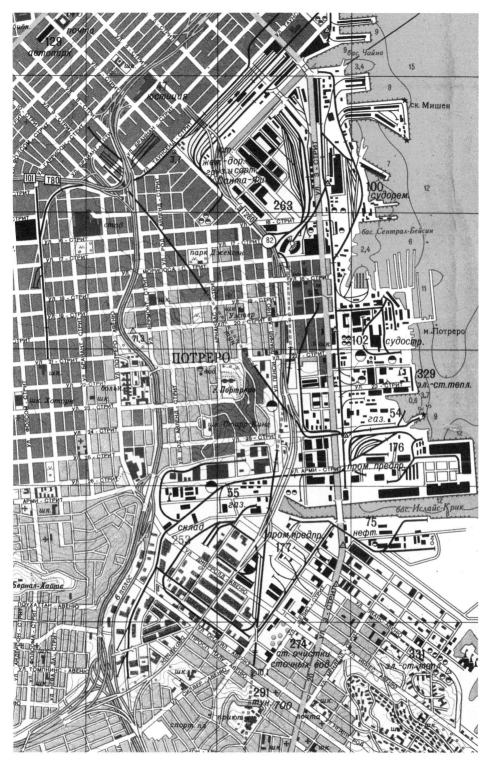

A1.29 *San Francisco*, 1:25,000, compiled 1976, printed 1980 (sheet 3 of 8). The road numbers 82 (*upper right*) and T80 (*upper left*) are incorrect.

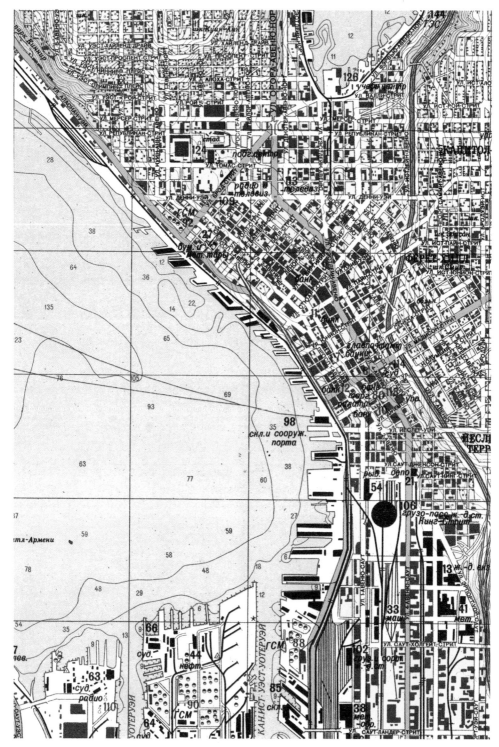

A1.30 *Seattle*, 1:25,000, compiled 1976, printed 1980 (sheet 2 of 3). Sea depths in Elliot Bay (*center* and *left*) are not shown on USGS mapping, and the values shown here do not correspond with the values on US Office of Coast Survey sheet 6442 of 1974.

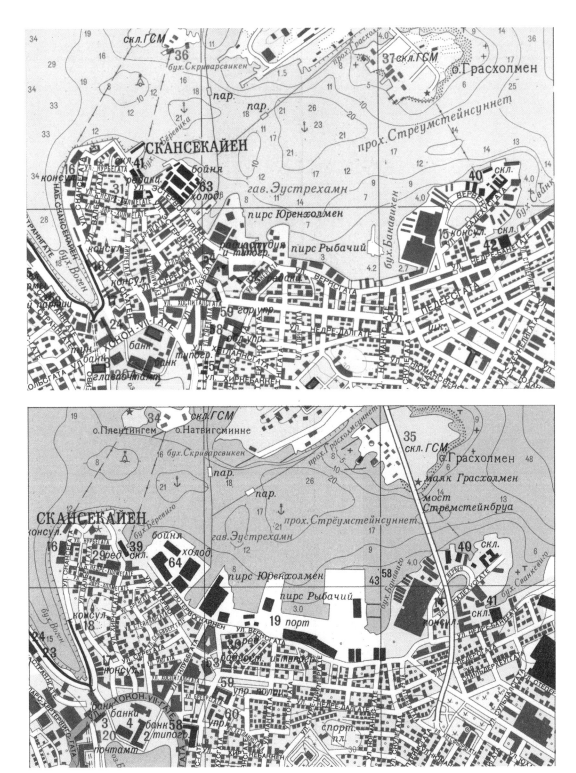

A1.31A and A1.31B *Stavanger*, Norway, 1:10,000, compiled 1971, printed 1975 (*above*) and compiled 1985, printed 1989 (*below*). Many differences can be seen in the depiction of both land and sea features.

A1.32 *Tokyo*, 1:20,000, compiled 1966, printed 1966 (sheet 2 of 4). A non-standard plan, with an unusual scale and style. The light-yellow shading indicates a densely built-up area, roads are uncolored, and railways and all labels are brown. Object 37 (*lower left*) is labeled "Imperial Palace."

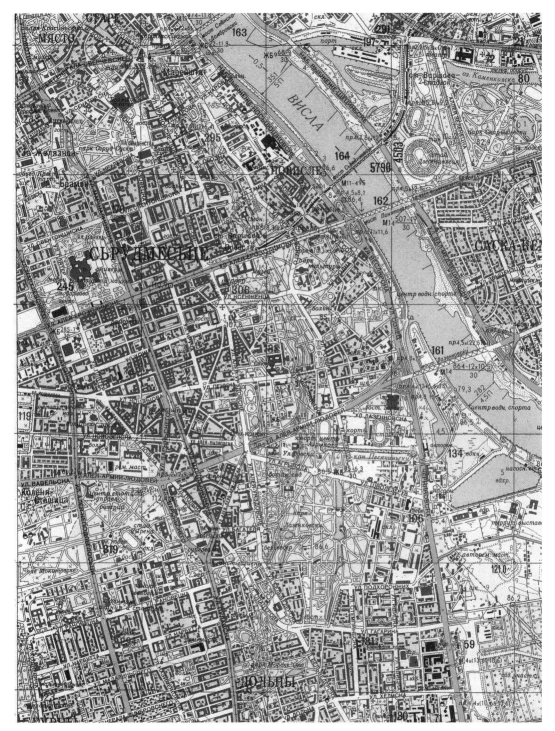

A1.33 *Warsaw*, 1:25,000, compiled 1980, printed 1981 (sheet 4 of 4). This plan of a Communist-era city is profusely annotated with the values of the characteristics of features. The river Vistula (*center top*) is noted as flowing north at a speed of 0.5 meters per second, with a width of 351 meters and a depth of 5 meters with a sandy bottom. The dimensions of the underpasses at each of the road and rail bridges are shown.

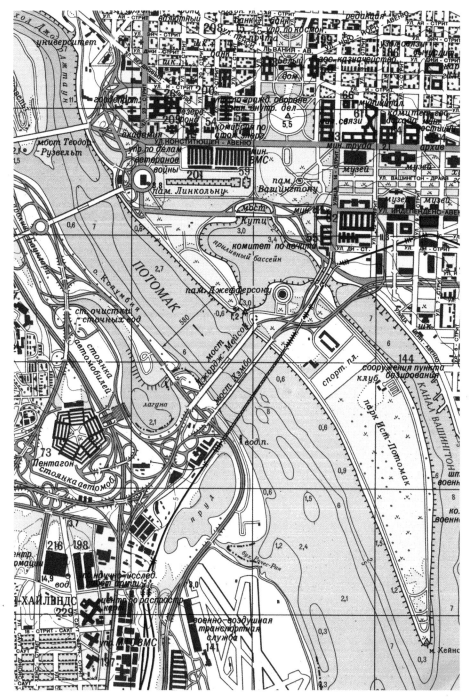

A1.34 *Washington, DC*, 1:25,000, compiled 1973, printed 1975 (sheet 3 of 4). Object 73 (*lower left*) is comprehensively labeled "Pentagon (Department of Defense, Army, Air Force and the Joint Chiefs of Staff)." At the upper center, objects 201 (*purple*) and 59 (*green*) are labeled "Department of Veterans Affairs" and "Department of the Navy," respectively (purple indicates an administrative establishment; green, military). These were temporary buildings erected during World Wars I and II and demolished in 1970, several years before the compilation of the map. A non-existent road, named "ВАШИНГТОН-ДРАВ" (Washington Drive), is shown running along the center of the National Mall.

A1.35 *Winston-Salem*, North Carolina, 1:25,000, compiled 1973, printed 1976 (sheet 1 of 1). In a rare instance of carelessness, a highway number is missing from the empty box (*center right*).

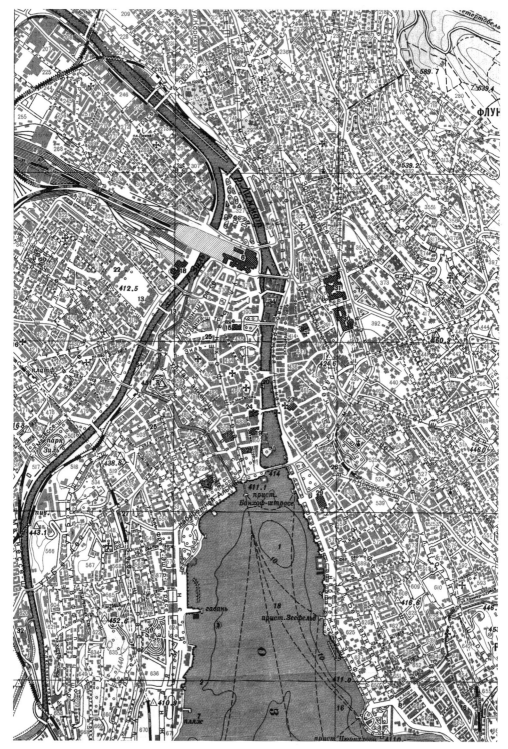

A1.36 *Zurich*, 1:15,000, compiled 1952, printed 1952 (sheet 1 of 1). An early plan with an unusual scale and style. The map has a list of objects, but no spravka or street index. As with most of the early maps, every plot of land is numbered (in pale brown).

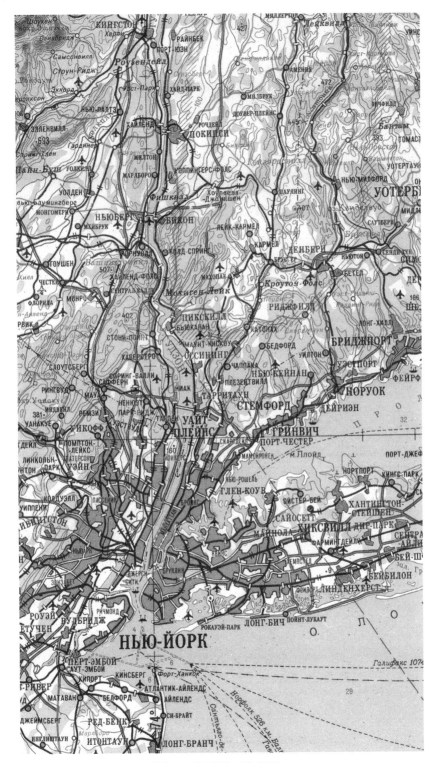

A1.37 1:1,000,000 sheet K-18, *New York*, compiled 1966, printed 1978.

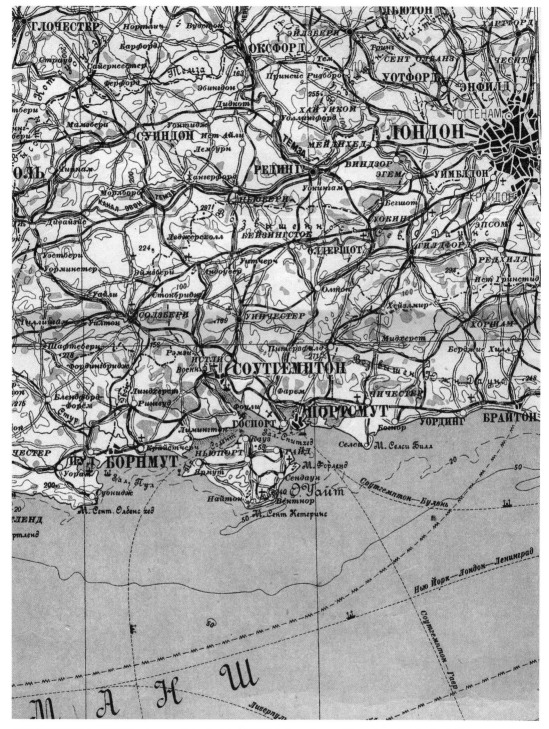

A1.38 1:1,000,000 sheet M-30, *London*, compiled and printed 1938. Note the undersea cable coming ashore (*center right*).

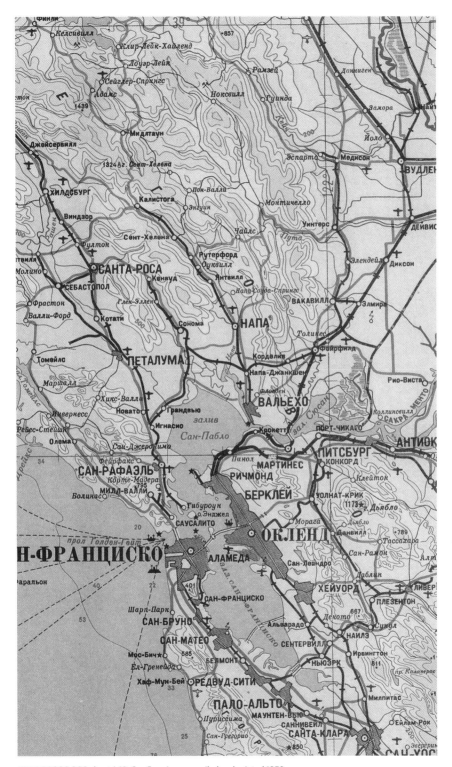

A1.39 1:1,000,000 sheet J-10, *San Francisco*, compiled and printed 1959.

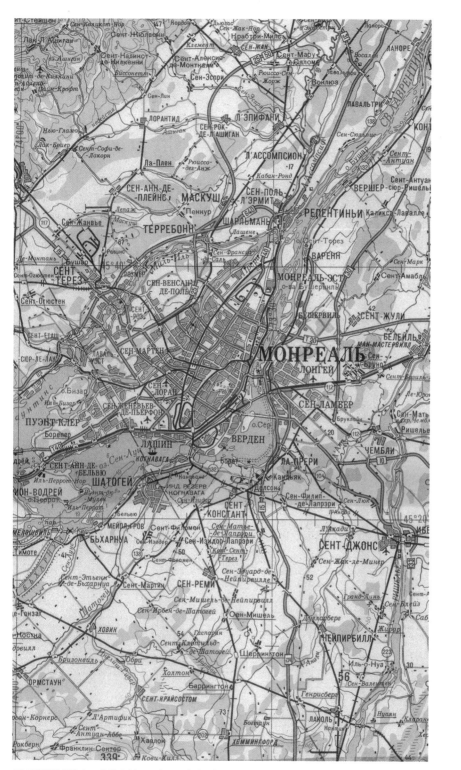

A1.40 1:500,000 sheet L-18-4, *Montreal*, compiled and printed 1981.

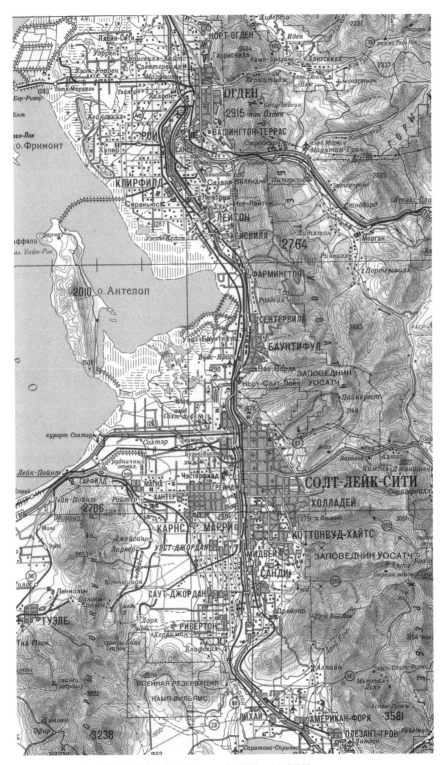

A1.41 1:500,000 sheet K-12-3, *Salt Lake City*, Utah, compiled 1980, printed 1982.

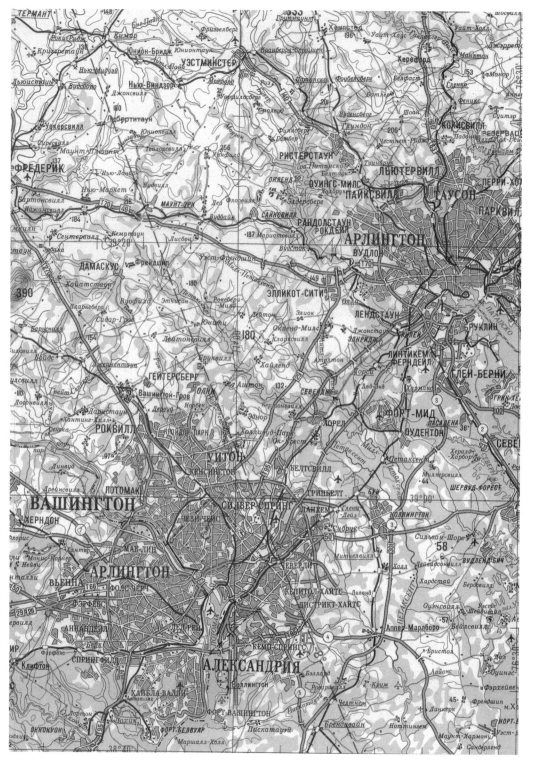

A1.42 1:500,000 sheet J-18-1, *Washington, DC*, compiled 1979, printed 1981.

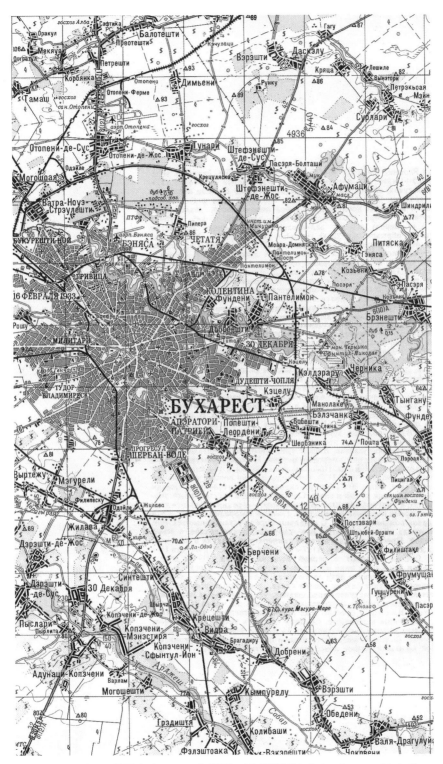

A1.43 1:200,000 sheet L-35-33, *Bucharest*, compiled 1975, printed 1985. Note that roads are annotated with distances, as well as with width and surface material.

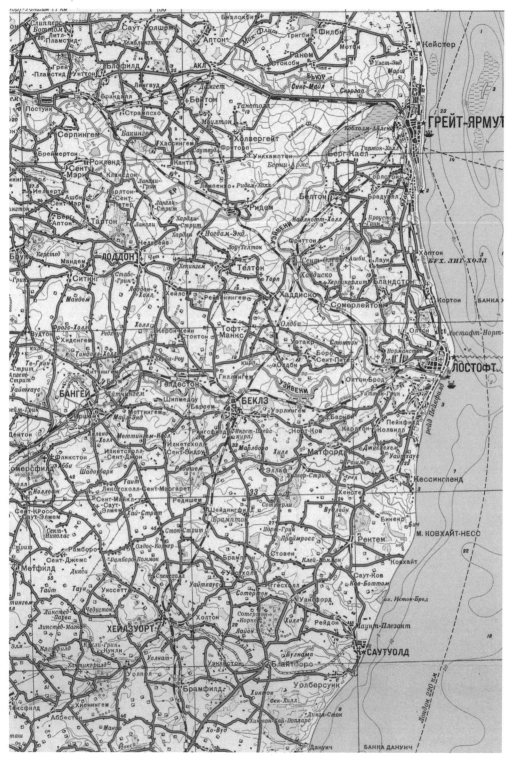

A1.44 1:200,000 sheet N-31-32 *Norwich*, UK, compiled 1949, printed 1950, showing parts of Norfolk and Suffolk. Note that the main and secondary roads are not differentiated, making through routes impossible to identify.

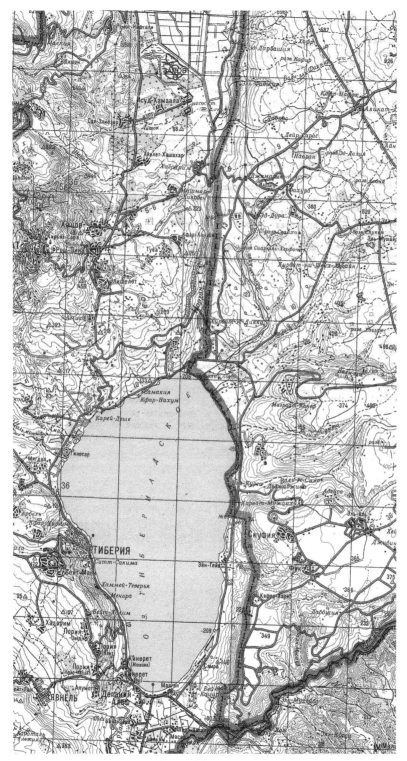

A1.45 1:200,000 sheet I-36-30 *Nazareth*, compiled 1983, printed 1985, showing the Sea of Galilee and parts of Israel, Jordan, and Syria.

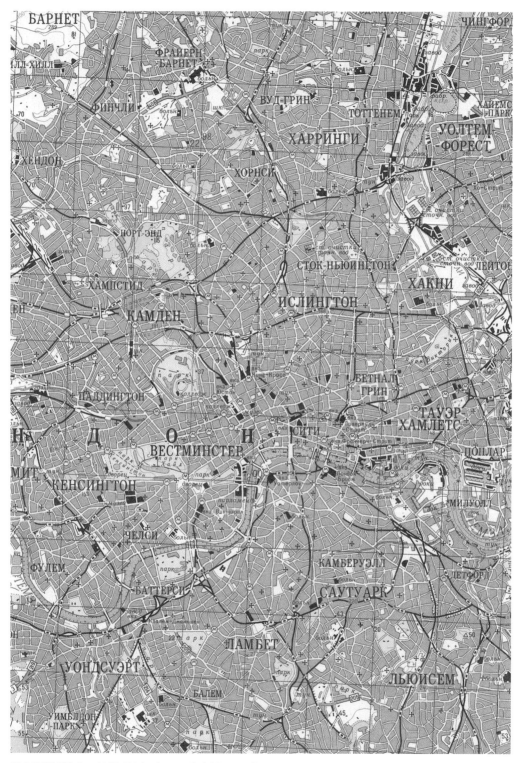

A1.46 1:100,000 sheet M-30-024, *London*, compiled 1981, printed 1983.

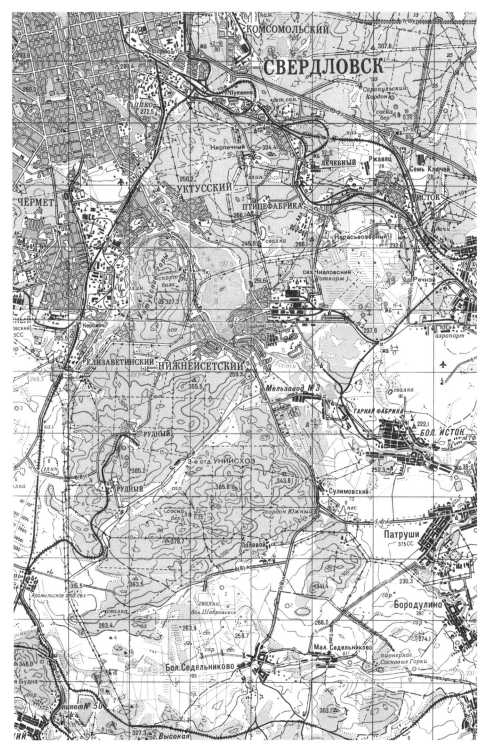

A1.47 1:100,000 SK-63 projection sheet W-18-31, untitled, printed 1985, produced by the GUGK showing Sverdlovsk (now Ekaterinburg), Russia. This map, produced for use by civil authorities, has no identifying geographic coordinates. This sheet approximates to SK-42 sheet O-41-110, although the sheet boundaries differ.

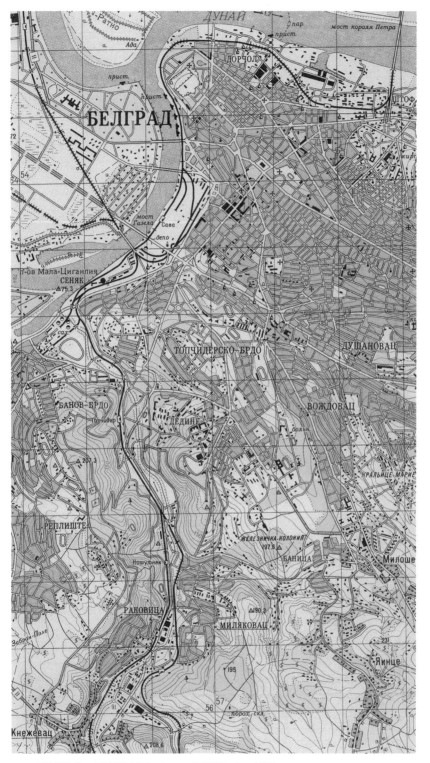

A1.48 1:50,000 sheet L-34-113-4, *Belgrade*, compiled 1971, printed 1973.

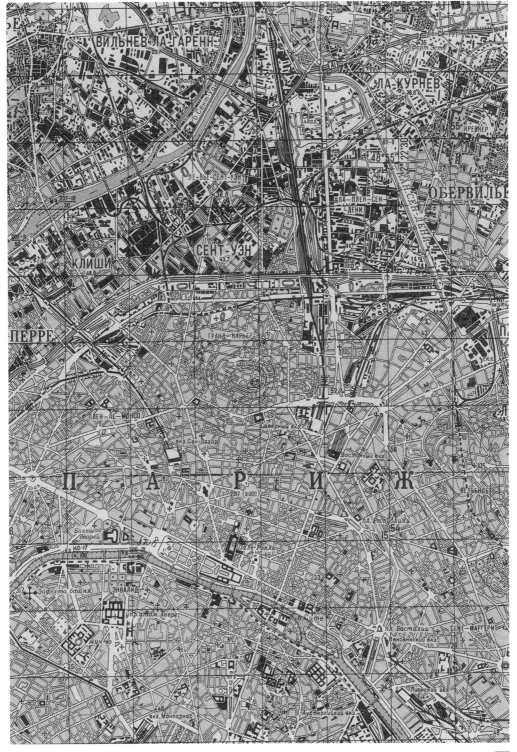

A1.49 1:50,000 sheet M-31-113-2, *Paris*, compiled and printed 1987.

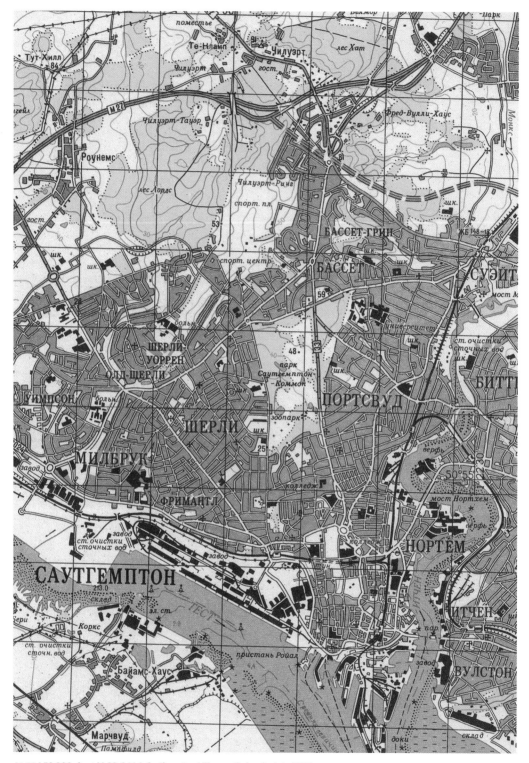

A1.50 1:50,000 sheet M-30-046-1, *Southampton*, UK, compiled and printed 1981.

A1.51 1:25,000 sheet O-35-097-4-2, *Gauja*, Latvia, showing the Ādaži military training ground, northeast of Riga, Latvia. The red overprinting marks the range boundary and battlefield features.

A1.52 1:25,000 sheet O-35-109-1-1, *Riga West*, Latvia, compiled and printed 1965. Unlike the city plans of the same scale, no street names appear on the topographic sheets.

A1.53 1:25,000 sheet O-36-001-2-1, *Leningrad*, 1941 edition.

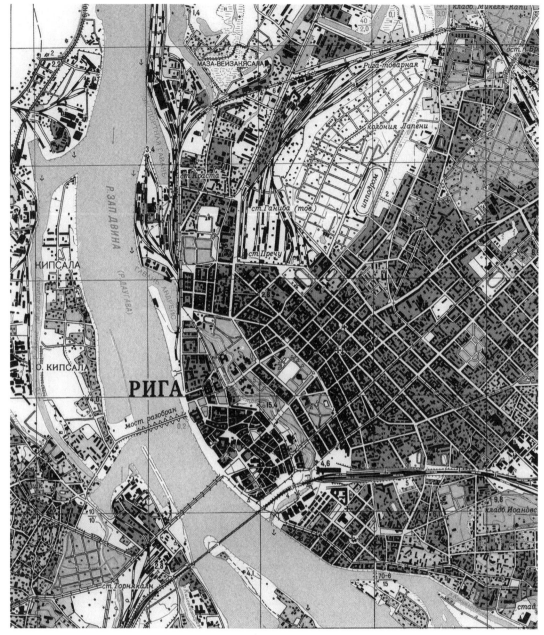

A1.54 1:25,000 SK-63 projection, sheet C-51-022-1-2, untitled, showing part of Riga, Latvia, printed 1987. The style and detail can be seen to be similar to the SK-42 version in fig A1.52.

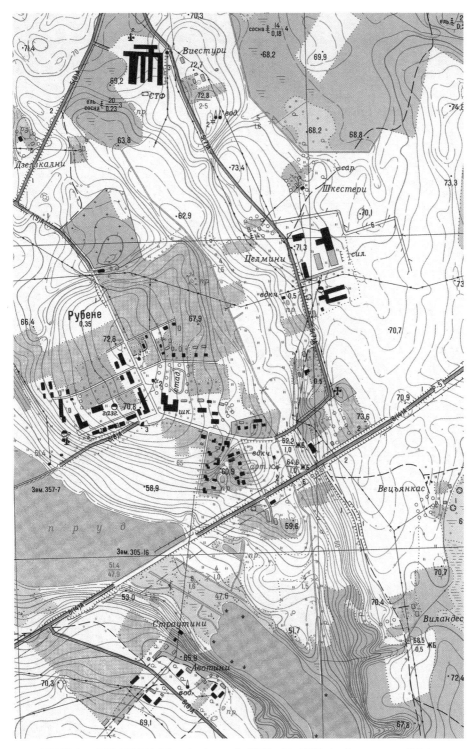

A1.55 1:10,000 sheet O-35-087-4-1-1, *Rubene*, Latvia, compiled 1986, printed 1988, showing an area near Valmiera, Latvia. The map has extensive annotation of the dimensions and characteristics of features such as roads, woods, waterways, and embankments.

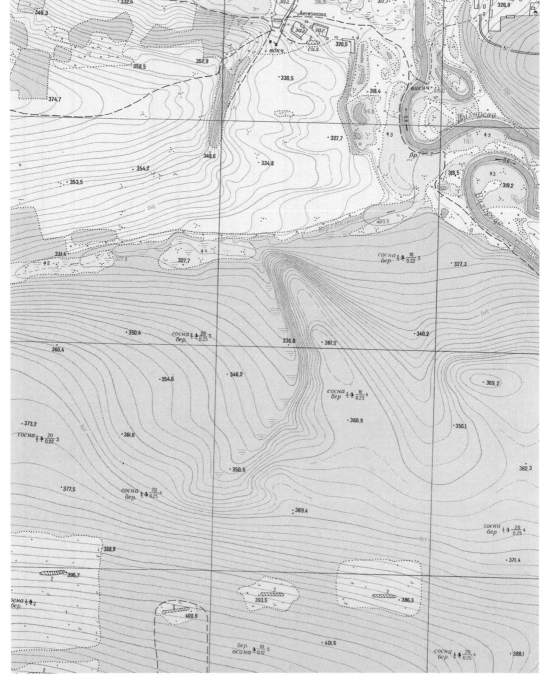

A1.56 1:10,000 SK-63 sheet F-36-031-2-4-2, untitled, showing unidentified part of Irkutsk Oblast, printed 1981.

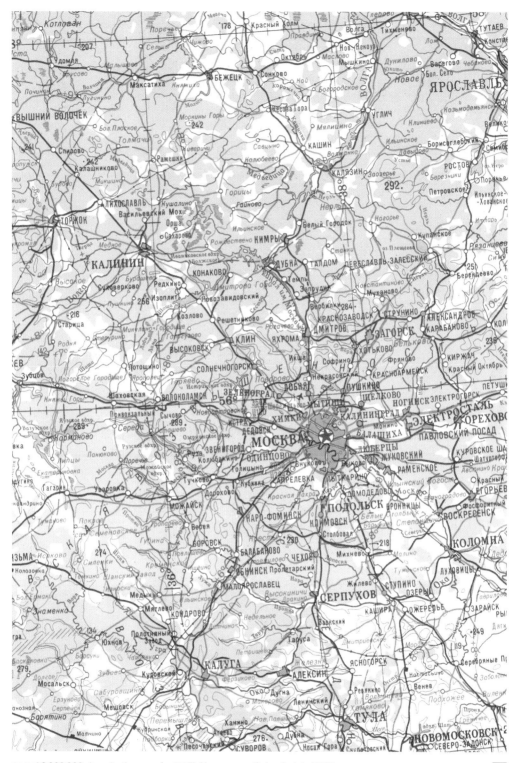

A1.57 1:2,000,000 air navigation map sheet Б-III, *Moscow*, compiled and printed 1985.

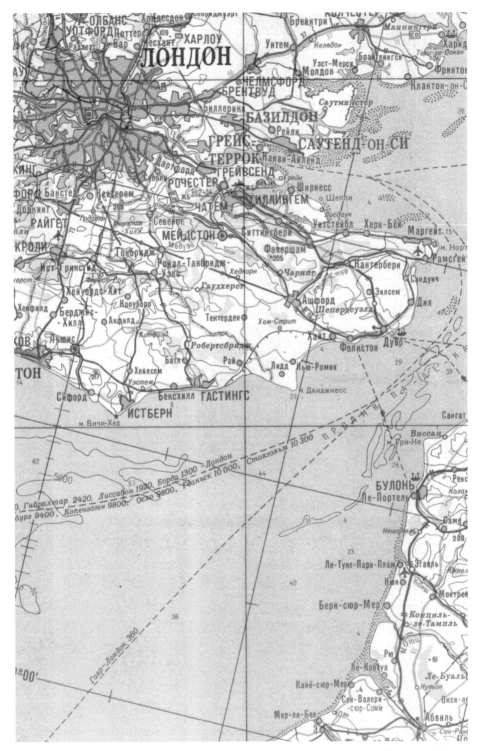

A1.58 1:1,000,000 rectangular series sheet 14-00-68, *London-Paris*, compiled and printed 1974. Unlike the standard topographic series, adjacent sheets of this series can be trimmed of margins and mounted together to give continuous coverage.

APPENDIX 2:
REFERENCES AND RESOURCES

IN ENGLISH:

1. *Conventional Signs and Abbreviations Used on USSR Military and Geodesy and Cartography Committee Maps*, Series GSGS 5861 (Military Survey [UK], 1992).

2. "Copying Maps Costs AA £20m," *Guardian*, March 6, 2001.

3. *Foreign Maps*, US Department of the Army Technical Manual TM 5-248 (1963). Includes a useful summary of USSR mapping activities, authorities, and map characteristics at that time.

4. *Glossary of Soviet Military and Related Abbreviations*, US Department of the Army Technical Manual TM 30-546 (1957).

5. *Red Army Maps of UK and Other Countries* (catalog) (Kerry, Wales: David Archer Maps, 1996).

6. "Russia Jails 'Spy' for Handing Maps to US Intelligence," *BBC News*, May 31, 2012.

7. *Russian Military Mapping: A Guide to Using the Most Comprehensive Source of Global Geospatial Intelligence* (Minneapolis: East View Press, 2005). This is a translation of the 2003 edition of the manual used by the Russian armed forces.

8. "Russian Spy Jailed for Sending Secret Army Maps to US," *BBC News*, May 13, 2010.

9. *Soviet Topographic Map Symbols*, US Department of the Army Technical Manual TM 30-548 (1958).

10. *Specifications for Topographic Map in Scale 1:50,000*, 2nd ed. (Riga: State Land Service of the Republic of Latvia Cartography Board, 2000).

11. *Symbols on Land Maps, Aeronautical and Special Naval Charts, Standardization Agreement (STANAG)* (Military Agency for Standardization, North Atlantic Treaty Organization [NATO], 2000). Unclassified.

12. *Terrain Analysis of Afghanistan* (Minneapolis: East View Press, 2003). Translations of the topographic descriptions on the Soviet 1:200,000 sheets.

13. *Terrain Analysis of Syria and Lebanon* (Minneapolis: East View Press, 2015).

14. *Terrain Analysis of Ukraine* (Minneapolis: East View Press, 2014).

15. "UK Government's Secret List of 'Probable Nuclear Targets' in 1970s Released," *Guardian*, June 5, 2014.

16. "USSR Planned to Invade Sweden," *Pravda Online*, February 21, 2003, www.pravdareport.com/news/russia/21-02-2003/21952-0.

17. "Where to Purchase Soviet Military Mapping," Information Sheet 1C, Cambridge University Library Map Department (UK), first issued September 7, 2001, and updated regularly; latest version dated December 8, 2015, is at www.lib.cam.ac.uk/deptserv/maps/1CBUYSOV.DOC.

18. P. Collier, D. Fontana, A. Pearson, and A. Ryder, "The State of Mapping in the Former Satellite Countries of Eastern Europe," *Cartographic Journal* 33, no. 2 (1996): 131–39.

19. John L. Cruickshank, "Mapping for a Multi-Lingual Military Alliance: The Case of East Germany," *The Ranger* [journal of the Defence Surveyors' Association, UK] (Winter 2009): 33–36.

20. John L. Cruickshank, "Military Mapping by Russia and the Soviet Union," in *The History of Cartography*, Vol. 6: *Cartography in the Twentieth Century*, ed. Mark Monmonier (Chicago: University of Chicago Press, 2015), 932–42.

21. A. J. Kent and P. Vujakovic, "Stylistic Diversity in European State 1:50,000 Topographic Maps," *Cartographic Journal* 46, no. 3 (2009): 179–213.

22. Nikolay N. Komedchikov [Institute of Geography, Russian Academy of Sciences, Moscow], "Copyright on Cartographic Works in the Russian Federation," *ACTA Scientiarum Polonorum, Geodesia et Descriptio Terrarum* 6, no. 3 (2007): 15–18.

23. Nikolay N. Komedchikov, "The General Theory of Cartography Under the Aspect of Semiotics," *Trans Internet Zeitschrift für Kulturwissenschaften* [Trans Internet journal for cultural studies], no. 16 (2005), http://www.inst.at/trans/16Nr/07_6/komedchikov.

24. Greg Miller, *Inside the Secret World of Russia's Cold War Map Makers*, Wired.com, 2015, http://www.wired.com/2015/07/secret-cold-war-maps.

25. Clifford J. Mugnier, "Grids & Datums: Republic of Estonia," *Photogrammetric Engineering & Remote Sensing* (August 2007): 869-70. Describes the 1963 Projection of Soviet maps for civil use.

26. Béla Pokoly, ed., *Cartography in Hungary 2003-2007* (Moscow: Proceedings of 14th General Assembly, August 4-9, 2007).

27. Alexey V. Postnikov, "Maps for Ordinary Consumers versus Maps for the Military: Double Standards of Map Accuracy in Soviet Cartography, 1917-1991," *Cartography and Geographic Information Science* 29, no. 3 (2002): 243-60.

28. Alexey V. Postnikov, *Russia in Maps: A History of the Geographical Study and Cartography of the Country* (Moscow: Nash Dom-L'Age D'Homme, 1996). Part of the Russia's Cultural Heritage series from the Russian State Library Collection.

29. Roskartografia (Russian State Mapping Service), *Toponymic Data Files: Automated Data Processing Systems: Development of Russia's National Catalogue of Geographic Names* (New York: United Nations Economic and Social Council, Seventh United Nations Conference on the Standardization of Geographical Names, 1998).

30. Michael Stankiewicz et al., *The Evolution of Mathematical Bases of Polish Topographic Maps During the Recent 80 Years* (Moscow: Proceedings of 23rd International Cartographic Conference, August 4-10, 2007).

31. Desmond Travers, *Soviet Military Mapping of Ireland During the Cold War* (Zurich: Parallel History Project on Co-operative Security [PHP], [n.d.]), http://www.php.isn.ethz.ch/lory1.ethz.ch/publications/areastudies/sovmilmap.html.

32. Dagmar Unverhau, ed., *State Security and Mapping in the German Democratic Republic: Map Falsification as a Consequence of Excessive Secrecy?* (Berlin: Lit Verlag, 2006).

IN FINNISH:

33. Erkki-Sakari Harju, *Suomen sotilaskartoitus, 400 vuotta* [Finnish military mapping, 400 years] (Helsinki: AtlasArt Oy, 2016).

IN GERMAN:

34. *Militärtopographie Lehrbuch für Offiziere* [Military topography textbook for officers] (Berlin: Verlag Des Ministeriums Für Nationale Verteidigung, 1960). Includes examples and symbology of DDR maps and those of West Germany, NATO, France, Britain, and the United States.

35. Gerhard L. Fasching, ed., *Militärisches Geowesen der DDR von den Anfängen*

bis zur Wiedervereinigung [East German military-topographic service from inception to unification] (Wien: Bundesministerium für Landesverteidigung, 2006). A history of the service with illustrations and map extracts.

IN POLISH:

36. *Wokowach Radzieckiej Doktryny Politycznej* [In the shackles of Soviet political doctrine] (Warsaw: Wydawca Geodeta, 2010). The history of the Polish Military-Topographic Service, 1945–90.

IN RUSSIAN:

37. *Fundamental Regulations for the Making of Topographic Maps at the Scales of 1:10,000, 1:25,000, 1:50,000 and 1:100,000* (Moscow: Head of the Military Topographic Directorate of the General Staff and the Head of the Main Administration for Geodesy and Cartography of the Ministry of Internal Affairs [MVD] of the USSR, Editorial-Publishing Department of the Military Topographic Service Moscow, 1956). There is also a 1984 edition.

38. *Handbook on Cartographic and Map-Issuing Works; Part 4: Compilation and Preparation for Printing of Plans of Towns* (Moscow: Chief of the Military-Topographic Directorate of the General Staff and by the Chief of the Main Administration of Geodesy and Cartography under the Council of Ministers of the USSR, 1978).

39. *Карта Офицера* [Map officer] (Moscow: General Directorate of Combat Training of the Ground Forces, 1985).

40. *Symbols on Topographic Maps 1:10,000 Scale* (Moscow: Head of Geodesy and Cartography under the Council of Ministers and the Head of the Military Topographic Directorate of the General Staff, 1970).

41. T. V. Vereshchaka, *Топографические Карты, Научные Основы Содержания* [Topographic maps: the scientific foundations of their content] (Moscow: MAIK "Nauka/Interperiodika," 2002).

IN SWEDISH:

42. Walther Blaadh (pseud.), *Sovjetisk Invasion av Sverige: Hur planerade Sovjet att invadera Sverige? Vad visste de? Hemliga kartor, planer och förband* [Soviet invasion of Sweden: How the Soviet Union planned to invade Sweden? What did they know? Secret maps, plans and formations], ed. Simon Olsson (Stockholm: Swedish Association for Military History, 2015).

43. Joakim von Braun and Lars Gyllenhall, *Ryska elitförband* [Russian elite forces] (Stockholm: Förlag Fischer, 2013). Includes Soviet military mapping of Sweden.

The following articles have appeared in *Sheetlines*, the journal of the Charles Close Society for the Study of Ordnance Survey maps, www.charlesclosesociety.org:

44. John Cruickshank, "German-Soviet Friendship and the Warsaw Pact Mapping of Britain and Western Europe," *Sheetlines* 79 (August 2007): 23-43.

45. John Cruickshank, "Виды из Москвы—Views from Moscow," *Sheetlines* 82 (August 2008): 37-49.

46. John Cruickshank, "Khrushchev Preferred Bartholomew's Maps," *Sheetlines* 87 (April 2010): 31-34.

47. John Cruickshank, "How Big a Map Does It Take to Build Socialism?," *Sheetlines* 89 (December 2010): 5-12.

48. John Davies, "Uncle Joe Knew Where You Lived: Soviet Mapping of Britain," *Sheetlines* 72 (April 2005): 26-38; and *Sheetlines* 73 (August 2005): 6-20.

49. John Davies, "Comrade Baranow, the Bouncing Czech, Penkilan Head and the World Map," *Sheetlines* 78 (April 2007): 32-33.

50. David Watt, "Soviet Military Mapping," *Sheetlines* 74 (December 2005): 9-12.

OTHER USEFUL RESOURCES:

51. The website http://redatlasbook.com has links to many of the items listed above and has other useful resources such as map extracts and detailed listings. It also has a copy of the Ordnance Survey statement of September 1997.

52. The following libraries have collections of Soviet maps for inspection on site:

Bodleian Library, Oxford, UK
British Library, London, UK
Cambridge University Library, Cambridge, UK
Library of Congress, Washington, DC
Library of Trinity College, Dublin, Ireland
National Library of Latvia

53. The following libraries have collections of Soviet maps viewable online:

ICGC Institut, Cartogràfic i Geològic de Catalunya: http://cartotecadigital.icc.cat/cdm/search/searchterm/govern%20sovietic/order/datea/lang/en_US

National Library of Australia: www.nla.gov.au

Stanford University: http://library.stanford.edu/guides/soviet-military-topographic-map-sets

54. Original paper copies of Soviet maps are on sale at Jana Seta map shop in Riga, Latvia: https://www.karsuveikals.lv/en

55. Digital images of Soviet maps are available as free download or for purchase from the following:

http://geospatial.com
http://www.landinfo.com
https://mapstor.com
http://maps.vlasenko.net
http://www.omnimap.com

APPENDIX 3:
TRANSLATION OF TYPICAL CITY PLAN "SPRAVKA"

TRANSLATION BY CHARLES AYLMER OF "INFORMATION" (СПРАВКА—spravka) from the *Cambridge*, UK, 1:10,000 City Plan of 1998.

GENERAL INFORMATION

Cambridge is a city in the southeast of Great Britain, the administrative center of the county of Cambridgeshire, and one of the country's oldest university centers. It is situated on the River Cam (actually a tributary of the River Ouse), 75 km. north of London. There are 90,000 inhabitants (1981); the area of the city is 25 km².

The city is surrounded by cultivated hills or flat plain (absolute height up to 70 m.), traversed by rather shallow river valleys. The altitude of the hills and ridges is 10–50 m., the summits of the hills are rounded or flat with gentle slopes. The predominant soil in Cambridge is boulder clay, with large areas of sand to the north of the city. When wet, the clay becomes waterlogged and severely impedes off-road movement of mechanized transport. The most important water obstacle in the city and its surroundings is the River Cam (or Granta), which is navigable north of Cambridge. Within the city limits the river is 10–30 m. wide and 1.5–2 m. deep; its banks are generally low and gently-sloping, and within the city reinforced in places with masonry. The other rivers are small (up to 10 m. wide). The rivers do not freeze and are in full flow all year round; high-water level is between November and February. Practically all land surrounding the city is under cultivation, with crops of wheat, barley, potatoes and sugar beet. Market gardens occupy a significant area, and there are orchards. Fields are bordered and roads lined with high hedges, which significantly impede observation of the countryside. A dense network of motor highways in the region of the city ensures the movement of transport in all directions throughout the year. The London-Cambridge motorway is dual carriageway with asphalt and ferro-concrete surfacing, each carriageway being 11 m. wide, with a dividing strip 2.5-5 m. wide. Improved highways have asphalt-concrete or asphalt surfacing; the width of the carriageway is 8-12 m. and the road-bed is 17-27 m. wide. The carriageway is edged with kerb-stones; there are specially prepared laybys—up to ten for every kilometer of road. Other highways are surfaced with asphalt, gravel or crushed stone; the width of the carriageway is 3-9 m., and the road-bed is 10-12 m. wide. Highway bridges are mainly of ferro-concrete or metal, with a load-bearing capacity of 60-80 t. The surroundings of the city are relatively sparsely inhabited. Rural inhabited locations have 50-500 inhabitants, the most important over 1000. The built-up area of important inhabited localities is in compact blocks or rows, that of small locations is widely spaced and irregular; there are many farms. Buildings are of 1-2 storeys and constructed of masonry. Small-holdings are usually bounded with hedgerows or enclosed by masonry walls. From the air Cambridge is easily recognized by its shape, its situation on the River Cam and the pattern of the road network.

The River Cam divides the city into two parts, which are connected by 18 bridges, including one railway bridge. The city does not have a uniform layout. Many areas of the city are rectilinear, with a distinct orientation, and some housing estates are almost radial or annular. The built-up area is predominantly dense or compact, but widely spaced in the outskirts. The old part of the city—its historic nucleus, most interesting architecturally—lies on a bend of the River Cam, on its right bank. It has characteristic narrow, often winding streets, with 2-4 storey masonry buildings of antique architecture (13th-19th centuries), including many beautiful mediaeval churches. Here are the Town Hall (objective 36) and the main post office (obj. 34). The bank of the River Cam is lined with ivy-clad buildings of the colleges of the university, with ridged roofs and turrets; all the buildings have large square inner courtyards with gates decorated with armorial bearings and bas-reliefs. The lodging-houses with their lecture-halls are reminiscent of monasteries or ancient castles. Beyond the old part of the city the buildings are of 2-5 storeys and of masonry construction; those on the outskirts of Cambridge are of 1-2 storeys, many of cottage type. In the blocks abutting the old part of the city to the south and south-east are a significant number of administrative and public establishments, including various agencies (obj. 50, 51 etc.) and the telephone exchange (obj. 45). The main streets of the city are wide (15-20 m.), minor roads are narrower (5-10 m.). Cambridge has plenty of greenery; there are many public spaces, parks and gardens. The principal industrial enterprises are concentrated in the northern and eastern outskirts. Cambridge is an important scientific center of Great Britain. The city contains a world-famous university—one of the most important elite universities in the country, which has its origins in the 13th century and retains many features of the Middle Ages in its organization. The university combines many colleges (obj. 3-7, 20-23) and carries out a broad range of scientific research. Each faculty of the university (including obj. 48) has its own particular speciality, but the number of subjects studied is small. The university has a number of scientific laboratories (obj. 24-27), museums, an art gallery and an important library (Z-5), which contains a large collection of valuable manuscripts. In Cambridge there are also a number of other institutes of higher education and scientific establishments, including a scientific research center (obj. 53), a pedagogical institute (obj. 18), a plant-breeding institute (obj. 19), the Scott polar institute, an observatory (obj. 28), scientific research and experimental stations (obj. 43 etc.).

Industrial and transport objectives

The main industry in Cambridge is mechanical engineering, principally electronic engineering and radio industry. There are important radio-electronics factories here (obj. 12-16), as well as instrument-making (obj. 11) and aircraft-maintenance (obj. 8) factories. Other factories produce building materials such as cement (obj. 17), concrete assemblies (obj. 9), asphalt, bricks and tiles, and there are many printing establishments. The food industry is well-developed. The Cambridge railway junction includes two stations (obj. 41 and 42), with comprehensive facilities and many warehouses, including storage for inflammable materials and lubricants (obj. 37).

Municipal economy, communications and medical facilities

Cambridge draws its electrical power from the national grid. The city has mains gas, produced by the local gas-works (obj. 10). There is running water (sourced from the Cam, 10 km. south of the city); two pumping stations (obj. 39 and 40) feed water through the city's distribution network. There is a sewerage system; sewage is disposed of in the river below the city, after processing at a special works (obj. 44). Buses are the principal means of transport in the city. The city has telephonic and telegraphic communication with many towns in Great Britain, and has an automatic telephone exchange (obj. 46). Cambridge has a broad range of medical and health facilities, including an important hospital (obj. 2).

APPENDIX 4:
TRANSLATION OF TYPICAL TOPOGRAPHIC MAP "SPRAVKA"

TRANSLATION BY CHARLES AYLMER OF "INFORMATION" (СПРАВКА—spravka) from verso side of 1:200,000 topographic map N-31-XXXI (N-31-31), *Cambridge*, UK.

LOCAL INFORMATION

Inhabited localities

Cambridge (90,000 inhabitants in 1981), the administrative center of the county of Cambridgeshire, is an important university center of Great Britain; it has well-developed book publishing and book trading industries. The towns of Swaffham, Downham Market, Chatteris and Hadley have 5–10 thousand inhabitants, the other towns 10–29 thousand. The towns are not laid out on a uniform plan (though certain parts of Cambridge are almost rectilinear). Built-up areas are compact in town centers, but widely-spread on the outskirts. Buildings are of masonry, mainly of 2–3 storeys. The majority of roads are narrow and winding, surfaced with asphalt or paving stones. Inhabited areas of rural type are typically farms and villages (50–500 inhabitants, the more important ones over 1000 inhabitants). Built-up areas of small inhabited localities of rural type are widely spaced and unsystematic; the more important ones are built-up in compact blocks or rows. Structures for the most part are of masonry and of 1–2 storeys. Small-holdings are usually bounded with hedgerows or masonry enclosures. Inhabited localities have mains electricity and telephone connections. The towns and many villages have running water; farms as [a] rule obtain their water from wells.

Almost all towns have mains gas. Railway tunnels (9220 and 6808) may be utilized as underground shelters. There are 23 airfields in the area, one of which is disused (7600).

Transport network

The main railway lines are twin-track; others are single-track; the gauge is 1435 mm. The London-Cambridge motorway is dual carriageway with ferro-concrete surfacing, each carriageway being 11 m. wide, with a dividing strip 2.5–5 m. wide. Improved highways (including European Route E112) have mainly asphalt-concrete or asphalt surfacing. The width of the carriageway is 8–12 m. and the road-bed is 17–27 m. wide. The carriageway is edged with kerb-stones; there are specially prepared laybys—up to ten for every kilometer of road. Other highways (including roads of local significance) are surfaced with asphalt, crushed stone or gravel. The width of the carriageway is 3–9 m., and the road-bed is 10–12 m. wide. The camber on improved highways is 4–6%, on other roads up to 7%. Bridges are mainly of ferro-concrete and masonry, with a load-bearing capacity of 60–80 t. (in individual cases up to 180 t.).

Relief and soils

The greater part of the area consists of a plain, which in the west is flat and traversed by rivers, canals and numerous drainage ditches, which significantly impede the off-road movement of wheeled vehicles; the east is hilly and in parts wooded. The southern part of the area is occupied by the East Anglian heights, divided by river valleys into ridges (absolute heights 80–130 m.), with flat or undulating crests, in places broken by ravines and hollows. River valleys within the area are for the most part broad and flat-bottomed with very gentle slopes, imperceptibly merging into the surrounding countryside; narrow valleys with steep slopes are found only in the southern part of the area. Soils in the western part of the plain are sandy (see diagram) to a depth of 6–20 m.; in the east are clay and loam to a depth of 4–15 m. The East Anglian heights are spread with boulder clay to a depth of 5–25 m. Ground water in the plain lies at a depth of 3–20 m., and 10–60 m. in the heights.

Hydrography

The River Ouse is navigable, 50–80 m. wide and 1.5–2 m. deep; its channel is canalized to a considerable extent. The rivers of Little Ouse, Nene, Cam, Lark, Well and Stour are navigable in their lower reaches; they are 10–30 m. wide and 1.5–2 m. deep; their channels have been canalized in certain sections. Other rivers are small (up to 20 m. wide). The banks of all rivers are mainly low and gently sloping; the flood-plain is cultivated and in the north-western part of the area is traversed by many irrigation ditches (up to 3 m. wide). The New Bedford, Old Bedford, Sixteen-Foot, Wisbech and Delph canals are navigable, 11–19 m. wide and 1.3–2.9 m. deep.

The rivers and canals are in full flow all year round; they do not freeze in winter. Peak water levels are from November to December. In the northern part of the area the water level of important rivers and canals varies periodically under the influence of sea tides.

Vegetation

Woods are mostly mixed (pine, oak, elm, beech), remarkably rarely coniferous (pine). Fields are bordered and roads lined with high hedges, which significantly impede observation of the countryside.

Climatic conditions

Winter (December–February) is mild, overcast and damp. Frosts are rare; the usual day air temperature is 5–7°C., at night 2–4°C. (minimum –14°C.). Precipitation mainly takes the form of incessant drizzle and occasionally wet snow, which usually melts immediately. There is precipitation on 12–16 days each month.

Spring (March–May) is protracted, with changeable weather. Usual day temperature is 10–16°C., but night frosts are possible until the end of April. There are 12–14 rainy days each month.

Summer (June–September) is moderately hot, with day temperatures of 16–18°C. (maximum 35°C., at night 11–13°C.). Precipitation takes the form of drizzle, with occasional downpours (sometimes thundery). There are 11–14 rainy days each month.

Autumn (October–November) is cool, with day temperatures usually

8–12°C., 4–8°C. at night; from the end of October regular night frosts begin. There is rain on 12–14 days each month.

Fog. In autumn and winter there are 2–4 foggy days each month (up to 6 days in October); the least number of foggy days (1–2 days) is in spring and summer. Prevailing winds throughout the year are south-westerly, southerly and northerly; average windspeed is 2–5 m/sec. In autumn and winter there are strong winds (with speeds of 15 m/sec and higher).

APPENDIX 5:
SYMBOLS AND ANNOTATION

This is a small selection of the symbology used on medium-scale and large-scale Soviet maps and plans.

COMMUNICATIONS AND TRANSPORTATION

A5.1 Airport

A5.2 Light railway or tramway

A5.3 Railway—double track electrified

A5.4 Railway—double track with station (main station building between tracks)

A5.5 Station—showing location of main building

A5.6 Disused railway

A5.7 Railway signal

A5.8 Railway overpass with height and width

A5.9 Railway bridge with material (M = metal), clearance under, length, and width

A5.10 Railway cutting with depth

A5.11 Railway embankment with height

A5.12 Railway tunnel with length, height, and width

A5.13 Divided highway with width of each lane, number of lanes, and surface material (Ц = concrete)

A5.14 Highway with width, clearance, and surface material (A = asphalt)

A5.16 Road numbers

A5.15 Minor road with width

A5.17 Ferry with width of stream, ferry dimensions, and load capacity

INDUSTRIAL AND URBAN

A5.18 Aerial ropeway

A5.25 Power station

A5.19 Cemetery

A5.26 Quarry with depth

A5.20 Electricity transmission line

A5.27 Urban areas—fire-resistant

A5.21 Gas holders

A5.28 Urban areas—other

A5.22 Important building, color-coded on map and listed in index

A5.29 Sakia (water-lifting device)

A5.23 Mine

A5.30 Windmill

A5.24 Oil well

A5.31 Wind pump

LANDSCAPE FEATURES ETC.

A5.32 Copse, open space with trees

A5.33 Lawn, grassy open space

A5.34 Single coniferous tree

A5.35 Single deciduous tree

A5.36 Mixed forest

A5.37 River—navigable
(name in uppercase)

A5.38 River—non-navigable
(name in lowercase)

A5.39 Tidal river—direction of incoming tide, direction of river flow

A5.40 Spot height

A5.41 Mound with height

A5.42 Monument

APPENDIX 6:
GLOSSARY OF COMMON TERMS AND ABBREVIATIONS

А	asphalt		паровоз	locomotive works
аэрп	airport		насос ст	pumping station
водопровод	water pipeline		пирс	pier/dock
больн	hospital		полиция	police
гараж	garage		почта	post office
ГСМ	fuel depot		пристань	pier/landing stage
гост	hotel		род	spring
депо	depot		спорт пл	sports field
ЖБ	reinforced concrete		скл	warehouse
завод	factory		ст	station
колледж	college		стад	stadium
М	metro station		суд	shipyard
метро	metro		ур	open space
мост	bridge		Ц	concrete (road surface)
тун	tunnel			
недейств	disused		шк	school
пар	ferry		заповедник	reservation/preserve
парк	park			

APPENDIX 7:
PRINT CODES

THE MAPS HAVE A PRINT CODE ON THE BOTTOM RIGHT-HAND corner, which is made up of the following: series code, job number/sheet, month (in Roman numerals) and year of printing, and factory code. For example, figure A7.1 shows the print code from the Miami city plan, sheet 2.

A7.1 Typical print code.

The code И-32 / 2 IV 84-Л indicates:

И . . . . Series—city plan
32 / 2 . . . . Job number 32, sheet 2
IV 84 . . . . Date of printing—April 1984
Л . . . . Factory—Leningrad

The series codes are as follows:

А . . . . 1:10,000 topographic map
Б . . . . 1:25,000 topographic map
В . . . . 1:50,000 topographic map
Г . . . . 1:100,000 topographic map
Д . . . . 1:200,000 topographic map
Е . . . . 1:500,000 topographic map
Ж . . . . 1:1,000,000 topographic map
З . . . . 1:2,000,000 and 1:4,000,000 aeronavigation map
И . . . . 1:10,000 and 1:25,000 city plan

Some 1:10,000 city plans of 1960s are coded A (e.g., see fig. 3.3).

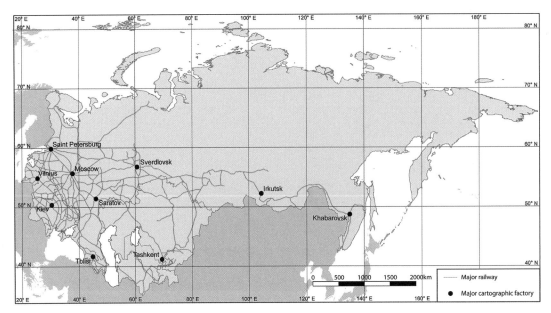

A7.2 Locations of known map factories.

Factory codes are as follows:

Б . . . . [not known]
В . . . . Vilnius [?]
Д . . . . Dunayev (Moscow)
Е . . . . [not known]
И . . . . Irkutsk
К . . . . Kiev
Л . . . . Leningrad (now St. Petersburg)
Ср . . . . Sverdlovsk (now Ekaterinburg)
Срт . . . . Saratov
Т . . . . Tashkent
Тб . . . . Tblisi
Хб . . . . Khabarovsk

APPENDIX 8:
SECRECY AND CONTROL

DURING SOVIET TIMES, THE OBSESSION WITH SECRECY EXTENDED beyond the contents of the maps to the very existence of the worldwide mapping project. Within the Soviet Union, even army officers who were required to use maps for training and exercises had no idea of the extent and scope of the project; all information was restricted on a "need to know" basis.

A8.1, A8.2, and A8.3 Red Army forms for the control of issuing and returning maps and managing the stocks.

Even access to individual map sheets was carefully controlled. Aivars Beldavs, now manager of the Jana Seta map shop in Riga, recalls his time as a Red Army officer: "Every map sheet needed for an exercise had to be signed out of the map store and signed back in. If it became damaged or destroyed during use, then even its remnants had to be returned."

PLACE-NAMES INDEX

Map extracts are indicated by page numbers in bold.

Aberdeen, UK, **58**
Afghanistan, 21, **134**, 134-35, 206
Alexandria, VA, **58**, 59
Almaty, Kazakhstan, **26**, 29, 42
Ape, Latvia, **44**
Armenia, 135
Astronaut Islands, CA, **121**, 122

Badajoz, Spain, 81
Baden-Baden, Germany, 48
Bagenalstown/Muine Bheag, Ireland, **82**
Bamian, Afghanistan, **134**
Barking, UK, **126**
Bayonne, NJ, 65, **68**, **73**
Beijing, China, 36, **147**
Belarus, 131, 142
Belfast, UK, 5, **6**, 48-49, 88
Belgrade, Serbia, **194**
Berlin, Germany, 43, 46, **138**, **148**, 207
Birkenshaw, UK, 93-97, **94**, **95**, **96**, 97
Birmingham, UK, 38, 117, **118**, 129
Boston, MA, 38, 53, **55**, 61, **64**, 99, 105, **106**, **107**, 114, 129-30, **149**
Bournemouth, UK, 33, **56**, 57, **59**, 92
Bradford, UK, 77, **78**, **79**, 80, **81**, 93, 94, 95, **96**, 97
Bristol, UK, **81**, 103, **150**

Bromborough, UK, 87, **88**
Bucharest, Romania, **189**
Budapest, Hungary, 46
Burghfield, UK, **60**
Burlington, MA, **55**

Cambridge, UK, 21, 38, **84**, **85**, 92, **116**, 133, **151**, 209, 211-18
Cameroon, 136
Cardiff, UK, **24**
Carlisle, UK, 32
Carlow, Ireland, **81**, 82, **137**
Caspian Sea, 42
Cēsis, Latvia, 132
Chambéry, France, **152**
Chatham, UK, 61, **62**, **63**, **72**, 129
Chicago, IL, 33, 46, 53, 107, **109**, **122**, **123**, **153**
Clyde (river), UK, 113, 126-27
Colchester, UK, **37**
Cologne, Germany, 132
Concord, MA, **65**
Copenhagen, Denmark, **154**
Cork, Ireland, 33
Cortez, CO, **12**
Crewe, UK, **52**, 53
Czechoslovakia, 25, 30

Darlington, UK, **35**, 36
Dewsbury, UK, 93, **95**, 97
Doncaster, UK, 83, **84**
Dortmund, Germany, 43
Dover, UK, 127
Dublin, Ireland, 36, 80–81, **111**, 112, 129–30, 210

Eastbourne, UK, **27**
East Germany, 21, 30, **138**, 139, 207–8
Edinburgh, UK, **61**, 73, **74**, 127, **155**
Ekaterinburg, Russia, **193**, 224
Enfield, UK, 118, **120**
Erskine, UK, **113**, 126

Falmouth, UK, 143
Finland, 5, 139–41, **158**, 207
Forth (river), UK, 127
Forth & Clyde Canal, UK, 126

Gainsborough, UK, 32
Gauja, Latvia, **197**
Gildersome, UK, 82
Gipton, UK, 82
Glasgow, UK, **83**, 103, **104**, **105**, **113**, 126–27, **156**
Gloucester, UK, 80
Goole, UK, 82
Granada Hills, CA, 53, **54**
Grand Junction, CO, **26**
Greenville, NJ, 65, **68**

Hague, The, Netherlands, 81
Halifax, Nova Scotia, **157**
Halifax, UK, **91**, 97
Hanscom Field, MA, 61, **64**, **65**, 76
Harlow, UK, **24**
Hastings, UK, **27**
Havant, UK, 97, **111**, 112
Havering-atte-Bower, UK, 118, **119**
Helsinki, Finland, **158**
Himalayas, 48
Huddersfield, UK, 85, **86**, 97, **117**

Hungary, 25, 30, 207

Irish Sea, 46, 164
Irkutsk Oblast, Russia, 29, **202**, 224
Israel, 136–37, **191**
Istanbul, Turkey, **159**

Jarrow, UK, **113**
Jēkabpils, Latvia, 39
Jersey City, NJ, **68**
Jordan, **191**

Karlskrona, Sweden, 70
Kazakhstan, **26**, 42
Kew, UK, 127
Keynsham, UK, **81**
Kiev, Ukraine, 49
Kilmarnock, UK, 49
Kingskerswell, UK, 118, **119**
Kingston upon Hull, UK, 32, 81–82
Krasnovishersk, Russia, 42
Kronstadt, Russia, 25

Lancaster, UK, 87, **88**
Latvia, 20, 21, **22**, 28, **29**, 38, 39, **40**, **41**, 42, **44**, 132, **137**, **138**, **197**, **198**, **200**, **201**, 205, 210
Lebanon, 21, 136–37, 206
Leeds, UK, 82, 93, **94**, **95**, **96**, 97, 103
Leicester, UK, 80
Leighlinbridge/Leithghlinn an Droichid, Ireland, **82**
Leningrad. *See* St. Petersburg/Leningrad, Russia
Lexington, MA, **55**
Limerick, Ireland, 33
Lithuania, **10**, 73–74, **75**, 76
Liverpool, UK, **31**, 87, **88**, **89**, 97, 105, 127, **160**
Ljubljana, Slovenia, 48, **161**
London, UK, 19, 31, **34–35**, 38, 43, 45, 73, **74**, 87, **90**, 91, 97, 101, **102**, 105, 107, **110**, **116**, 117, 118, **119**, **120**,

125, **126**, 129, 133, **162**, **184**, **192**, **204**, 209, 211-12, 216
Los Angeles, CA, 33, 38, **53**, **54**, **121**, 122, 129, **163**
Luton, UK, 33, 91, **92**, **93**

Manchester, UK, 83, 97, **164**
Margate, UK, **27**
Maribor, Slovenia, 48, **165**
Marseilles, France, 81
Matishi, Latvia, 20
Medway River, UK, **72**, 127
Mersey (river), UK, 87, **88**, 89, 127
Miami, FL, **56**, **57**, **72**, 73, 129, **166**, 223
Middlesbrough, UK, 92
Montreal, Quebec, **167**, **186**
Moscow, Russia, 43, 45, 49, 81, 140-41, **203**, 206-9, 224

Nazareth, Israel, **191**
Neilston, UK, 103, **104**
Nepal, 136
Newark, NJ, **73**
Newcastle upon Tyne, UK, 97, **113**, **169**
New York, NY, 21, 26, 31, 33, 65, **68**, **69**, **73**, 105, 107, **108**, **114**, **168**, **183**, 207
Nitshill, UK, **83**
Norwich, UK, **190**

Okehampton, UK, 20
Opa-Locka, FL, **56**, **57**
Ottawa, Ontario, 26, 46
Oxford, UK, 61, 129, **170**, 209

Paignton, UK, **98**, 99
Palestine, 136-37
Paris, France, **27**, 45, 140, **195**, **204**
Paterson, NJ, 113, **114**
Pembroke, UK, **50**, **51**, **52**, 83
Plymouth, UK, 24, 61, **63**
Point Loma, CA, **66**, **67**
Poland, 25, 30, 207
Portland, ME, **171**

Portsmouth, UK, 97, **111**, 112
Prague, Czech Republic, **172**
Pulkovo Observatory, Russia, 25

Raleigh, NC, **103**, **173**
Reading, UK, 60, 92
Riga, Latvia, 21-22, **22**, 39, 42, 132, **197**, **198**, **200**, 205, 210, 226
Romania, 30
Rotterdam, Netherlands, 105
Rubene, Latvia, **201**

Salt Lake City, UT, **187**
San Diego, CA, 61, **66**, **67**, 123, **124**, 129, **174**
San Fernando, CA, 53, **54**, **163**
San Francisco. CA, 33, **72**, 73, **99**, 101, 107, **109**, **112**, **175**, **185**
San Pedro Bay, CA, **121**, 122
Sari, Iran, 49
Seattle, WA, 129, **176**
Sheffield, UK, 101, **102**
Southampton, UK, **27**, **100**, 101, **125**, 128-29, **196**
Sowood Green, UK, 85, **86**
Stavanger, Norway, **177**
St. Helens, UK, 97, **114**, 115, 164
St. Louis, Missouri, 45, 46
St. Petersburg/Leningrad, Russia, 4, 25, 49, 140, **199**, 224
Stockholm, Sweden, 70-71, 105, 209
Suez, Egypt, 81
Sunderland. UK, 97
Sverdlovsk Oblast, Russia, 29, **30**, **193**, 224
Sweden, 70-71, 105, 142, 206, 208-9
Syria, 21, 136, **191**, 206

Tajikistan, 42
Tallinn, Estonia, 131-32
Tangiers, Morocco, 81
Tees (river), UK, 53, **54**, 127
Teesside, UK, **53**, 54

Tehran, Iran, 81
Thames (river), UK, 112, **126**
Thornaby-on-Tees, UK, 53, **54**
Thurrock, UK, 97
Tokyo, Japan, **178**
Torbay, UK, **98**, 100, **119**
Toronto, Ontario, 26
Torquay, UK, 118, **119**
Totley (tunnel), UK, **101**, **102**
Trondheim, Norway, 43
Turkmenistan, 42
Tyne (river), UK, **113**, 127, **169**

Ukraine, 21, 131, 206
Upper Heyford, UK, 61, **64**, 83
Uzbekistan, 29, 42

Valmiera, Latvia, 38, 39, **40**, **41**, **201**
Vancouver, British Columbia, 53, 143

Ventspils, Latvia, 39
Vilnius, Lithuania, **10**, 73, 74, **75**, 76, **224**

Warrington, UK, 97, 164
Warsaw, Poland, 25, 30, 46, 139, **179**, 208–9
Washington, DC, **58**, 59, 70, 107, **110**, 129, **180**, **188**, 209
Wembley, UK, 101, **102**
Westbourne, UK, **56**, 57
Wigan, UK, 97, **115**
Winston-Salem, NC, **181**
Wolverhampton, UK, 92, **118**
Wymondham, UK, 81

York, UK, 101, **102**

Zurich, Switzerland, 49, **182**, 207

GENERAL INDEX

AA (motoring organization), 133
Academy of Sciences research ships, 127
Admiralty chart, 120, 125-26
Aerogeodezija (North West Aerogeodetic Institute), 140
aeronavigation maps, 11, 42, 45-46, 223
Aftonbladet (newspaper), 70
Agency for International Development, 135
Akademik Kovalevskij (Soviet ship), 127
Alexander (tsar), 4
altitude. *See* spot heights
ArcGIS (software), 136
Archer, David, 133
areas of new housing, depiction of, 57-59
AV (Ausgabe für die Volkswirtschaft; East German Edition for the National Economy), 30, 139

Bartholomew maps, 88
bathometric depths, 120-26
Beldavs, Aivars, 226
bridges, depiction of, 72-74
British Library, 133

cable cars, non-depiction of, 112
Cambridge University Library, 133
captured German mapping, 49, 88-89
carrying capacity, annotation of, 72-74, 220

cartographers' handbooks, 13
cartographers' names on maps, 48-49
Central Administration for Geodesy and Cartography of the USSR Council of Ministers (GUGK), 9-10, 28, 30, 38-39, 42-43, 93
city plans, 30-41, 146-82
civil series, 38-41
contour lines, 36, 39, 76, 91, 172
"Control of Soviet Vessels in British Territorial Waters 1963-65," 127
coordinate system 1942 (SK-42), 19-27
coordinate system 1963 (SK-63), 28-30, 193, 200, 202
Crown copyright, 133
Cruickshank, John L., 206, 209
cultural expectations, 47, 87

Davies, John, 209
description of locality. *See* spravka
dredged channels, 120, 125-27

Edition for the National Economy (East German). *See* AV
elevations. *See* spot heights
EOTR (Hungarian local coordinate system), 30
errors on maps, 53-56

European road numbers (on British roads), 115–16

factories, identified on maps, 38, 129
factory codes. *See* print codes
Fairchild maps, 135
Fasching, Gerhard L., 208
ferries, depiction of, 112–13, 220
Finnish Ministry of Foreign Affairs, 140
forests, depiction of, 76, 137
Foshberg, Tore, 71
FSB (Federal Security Service), 142

Gaelic language place-names, 81–82
Gauss-Krüger, 25, 30, 33
General Staff, 9, 21
Geodata (company), 141
geological sketch map, 21–22
German mapping (captured), 49, 88
G-K. *See* Gauss-Krüger
Google Earth, 60
Google Maps, 60
graticules, 23, 25
GUGK. *See* Central Administration for Geodesy and Cartography of the USSR Council of Ministers
GUKiK (Polish local coordinate system), 30
Guy, Russell, 131–32

handbooks, cartographers', 13
Harju, Erkki-Sakari, 140
Her Majesty's Theatre, London, 87
Hezbollah-Israeli war, 136
high-rise buildings, depiction of, 100–101
highways, depiction of, 113–20, 219–20
Hitler, Adolf, 4–5
Holm, Christer, 70
hydro-aerodrome, 87–88
hydrography, 120–28
hyphenation of place-names, 83

Iceberg (Soviet ship), 127

important objects, list of, 30–31, 38–39, 128–29
IMW. *See* International Map of the World
International Cartographic Conference, 132
International Map of the World (IMW), 4, 8, 19, 46
Irving, E. G. (rear admiral), 127

Jana Seta map shop, Riga, 132, 210, 226
John F. Kennedy International Airport, 65, 69
Jolley, Craig, 135
JTSK (Czech local coordinate system), 30

Karta Mira. *See* World Map
Kaye, Thom, 135
Kelvin Hughes (navigation equipment), 128
Kent, A. J., 206
KGB, 71, 132
Khrushchev, Nikita, 209
Komedchikov, Nikolay N., 206

Latvian Geospatial Information Agency, 138–39
Lazar, Vladimir (colonel), 142
Lenin, Vladimir, 4
Lie of the Land (exhibition), 133
local coordinate system, 38–41
local pronunciation, 80–81
low-rise buildings, depiction of, 100–101

Mankiewicz, D. A. (colonel), 48
Mechanics' Institute, 85–86
metro systems, 105–10
MGD (Military Geographic Service of West Germany), 139
Mikhael Lomondsov (Soviet ship), 127
military series city plans, 30–38
military series topographic maps, 19–27
Military Topographic Depot, 4

Military Topographic Directorate of the General Staff of the Soviet Army (VTU), 9, 11, 28, 42, 143
Miller, Greg, 131, 207
misinterpretation of imagery, 53, 84
Molotov-Ribbentrop Pact (Pact of Steel), 5
MTD (East German Military Topographic Service), 139
Mugnier, Clifford J., 207

Napoléon, 4
National Archives, The (TNA), 127
National Geospatial Intelligence Agency, 135
New Zealand army, 134
nomenclature of maps. *See* numbering system
numbering system (of map sheets), 19–23, 29

officers' guides, 13
oil platform, 122–23
Okeanograf (Soviet ship), 127
Operation Barbarossa, 5
Operation Cast Lead, 136
Ordnance Survey (OS), 49–50, 53, 57, 60–61, 63–64, 76–77, 80–81, 84–85, 87, 89, 103–4, 113, 122, 126, 133
overlapping city plans, 92, 97

Pact of Steel (Molotov-Ribbentrop Pact), 5
Pask, Stuart, 135
Penck, Albrecht, 4
phonetic transliteration, 80–81
photogrammetry (aerial photography), 4, 9
place-names (toponyms), 80–83; hyphenation of, 83; pronunciation of, 80–81; underlined, 83
Planheft, 49
Pokoly, Béla, 207
Polyus (Soviet ship), 127
posters (for training purposes), 13–18

Postnikov, Alexey, 4, 207
Pravda Online, 70, 206
print codes, 49, 223
print quality of maps, 7
Priroda (company), 141
Pulkovo Observatory, 25

railroad company names, 103
railroads, depiction of, 101–10, 219; dismantled or disused, 7, 103–4, 143; electrified, 101–2, 219
rectangular topographic maps, 45–46, 204
redatlasbook.com, 209
remapping, 33, 91–92
rivers, navigable/non-navigable, depiction of, 13, 74, 221
roads, depiction of, 113–19, 219–20
Roman pottery kilns, 84
Roskartografia (Russian State Mapping Service), 207
Royal Naval Dockyard, 50–52, 61–62

satellite imagery, 7, 53, 61, 133, 135–36
scale (of maps), 3, 4, 5, 8, 9, 11, 19, 21, 28, 42
security classifications, 19–20
security omissions/deletions, 60–61
Shalman, I. I. (colonel), 48
Shiroky, Pyotr, 71
Sipachev, Gennady, 142
SK-42. *See* coordinate system 1942
SK-63. *See* coordinate system 1963
Sojuzkarta (company), 140
source materials (for cartographers), 47–48
Soviet ships visiting Britain, 127
spot heights, 76–79, 97
spravka, 30, 38, 211–18
Sputnik, 48
Stalin, Joseph, 3–5, 9, 143
Stereo-70 (Romanian local coordinate system), 30

streetcars, depiction of, 112
street index, 30-31, 38, 120
Stvor (Soviet ship), 127
Swedish naval base, 71
symbology, 11-18, 219-21

Taylor, Damon, 134
tidal range, 120-26
topographic maps (topos), 19-30, 42-46, 183-204
toponyms. *See* place-names
tourists, 9, 48
tramways, depiction of, 112
transliteration of place-names, 80-83
Transverse Mercator, 25
Travers, Desmond (ret. colonel), 136-37, 207
trunk roads, 116
tube network, 105, 107, 110

UK Hydrographic Office, 120, 125-26
underlined place-names, 83
unknown production, 129
Unverhau, Dagmar, 30, 207

US Air Force base, 64-65
US Army Technical Manual, 141, 205
US Geological Service (USGS), 53, 55, 57, 59, 61, 65, 76, 103, 112, 120, 122-23
US Marine Corps Recruiting Depot, 64
US Naval Reservation, 65, 68
US Naval Training Center, 64
US Navy Base, 61, 66
US Office of Coast Survey, 120-24

VTU. *See* Military Topographic Directorate of the General Staff of the Soviet army
Vujakovic, P., 206

Warsaw Pact, 25, 30, 46, 139, 209
Watt, David, 132, 209
world map (Karta Mira), 46

Yudin, A. D. (colonel), 48

Zarya (Soviet ship), 127
Zenit satellite, 7, 53
Zubov (Soviet ship), 127
Zvirbulis, Aivars, 132